Iroquois Corn in a Culture-Based Curriculum

SUNY series,
The Social Context of Education

Christine E. Sleeter, editor

Iroquois Corn in a Culture-Based Curriculum

A Framework for Respectfully Teaching about Cultures

Carol Cornelius

STATE UNIVERSITY OF NEW YORK PRESS

#38580212

Cover illustrated (Three Women Hoeing) and gallery illustrations by Ernest Smith Courtesy of Rochester Museum of Arts and Sciences, Rochester, NY.

Production by Ruth Fisher
Marketing by Fran Keneston

Published by
State University of New York Press, Albany

For information, address the State University of New York Press, State University Plaza, Albany, NY 12246

Library of Congress Cataloging-in-Publication Data

Cornelius, Carol, 1948–
    Iroquois corn in a culture-based curriculum : framework for respectfully teaching about cultures / Carol Cornelius.
        p.   cm. — (SUNY series, the social context of education)
    Includes bibliographical references (p.   ) and index.
    ISBN 0-7914-4027-3 (HC : acid free). — ISBN 0-7914-4028-1 (PB : acid free)
    1. Iroquois Indians—Study and teaching. 2. Multicultural education. 3. Indians of North America—Textbooks. 4. Stereotypes (Psychology). I. Title. II. Series: SUNY series, social context of education.
E99.I7C85   1999
370.117—dc21   98-14900
                                                              CIP

To my children,

family,

extended family,

and seven generations to come.

Table of Contents

List of Figures

Introduction

Why do American people have such distorted and stereotypical images of American Indians, or plead complete ignorance about American Indians? The answer to this question is quite easily understood when one realizes that the American people have not been taught about American Indians in a way that is accurate or respectful. The shortcomings of the educational process affect the multitude of diverse cultures within this nation because teachers have not been required to learn about diverse cultures except at a very surface level. There are a few, rare, fantastic teachers who have made a concerted effort to learn about diverse cultures and I applaud them. However, the vast majority have not been required, nor have they taken the time, to learn about diversity. This book provides a way to begin examining culture-based curriculum.

This text begins with an examination of the common stereotypes about American Indians that exist in academia, the media, and text-books. Studies show that ninety-five percent of what students know about American Indians was acquired through the media, and that teachers can erase stereotypes with accurate information.

Chapter 2 explores the academic theories behind stereotypes and provides insight on the creation of evolutionary theories and hierarchical scales that devalue indigenous cultures.

Chapter 3 illustrates how these theories became the standard curriculum, which explains why diversity has not been incorporated into textbooks.

Chapter 4 presents the theories that support the culture-based curriculum framework, which provides a new way to study diverse cultures in a respectful manner, and key elements for developing culture-based curriculum.

Chapters 5 through 9 provide the case study in which the thematic focus of corn has been identified and utilized as a central theme to study the Haudenosaunee (Iroquois) culture.

Terminology

Several terms used throughout this study need clarifying and defining from a culture-based perspective:

Culture

Culture has been defined from many perspectives in anthropology, sociology, education, and history. In this study culture has been defined as those indigenous peoples who have their own cosmology, world view, language, ceremonies, government, economic system, land base, health systems, and traditions that are rooted in antiquity. Indigenous peoples have a culturally specific way in which they perceive reality, that is, how they make meaning of this world, and that reality is based in ancient beliefs about how this world originated and how human beings should live on this earth.

Haudenosaunee

Haudenosaunee is translated into English as "People of the Longhouse." The Haudenosaunee are commonly known as the Iroquois Confederacy, which is the French label. The Five Nations, and later Six Nations, are labels by which the British identified the Haudenosaunee.

The Haudenosaunee consist of the six nations: Mohawk, Oneida, Onondaga, Cayuga, and Seneca, and the Tuscarora who were added in 1722. Even these names are misnomers, because in each of the six languages, the people have a name for themselves. An overall term that is often used to indicate indigenous people is *Ongwe'ho:weh* which translates to mean the "Real People." Each nation has a descriptive name:

> Mohawk—People of the Flint
> Oneida—People of the Standing Stone
> Onondaga—People of the Hills
> Cayuga—People of the Great Swamp or the Great Pipe
> Seneca—People of the Great Mountain
> Tuscarora—Shirt Wearing People

Each of the Haudenosaunee Nations lived in a definable territory as indicated in figure 1, Original Land Base, and continue to occupy small portions of their former land base as indicated in figure 2, Current Land Base. The Oneida Nation has three land bases (Wisconsin, Canada, and New York) and the Seneca/Cayuga have land in Oklahoma. The Cayuga are the only Nation which does not have a portion of their aboriginal land base in New York State. They live on the Cattaraugus Seneca Reservation in New York State, and have lands in Oklahoma and Canada.

"Indian"—"American Indian"—"Native American"

In this study the labels "American Indian," "Indian," and "Native American" are used interchangeably. The terms "American Indian" and "Indian" are labels that originated at the time of Columbus. He thought he had discovered the Indies so he labelled the peoples he found Indians. "Native Americans," as the more acceptable terminology, grew out of the civil rights protests of the 1960s–70s. Today, anyone born in this country tends to label themselves as native, therefore the term has lost significance in identifying the Indigenous Peoples of this hemisphere. Thus, in this study, "American Indian" and "Native American" are used interchangeably because they are irrelevant terms. Specific Indigenous cultures are designated with the name by which they identify themselves.

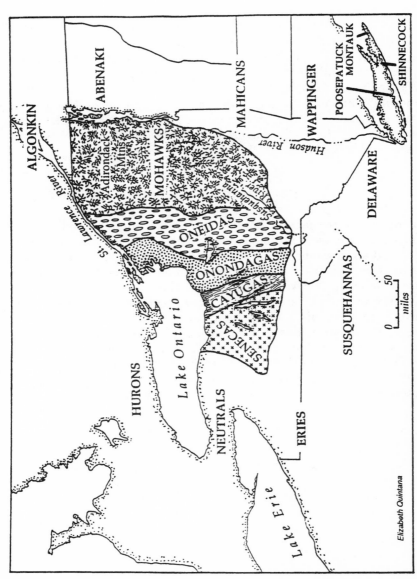

Figure 1 Lands of the Haudenosaunee—Original Land Base

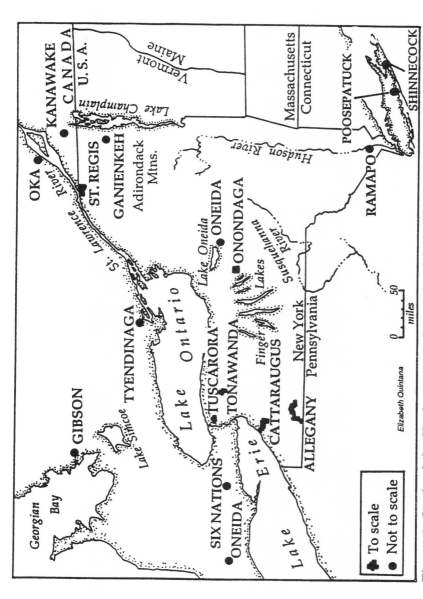

Figure 2 Lands of the Haudenosaunee—Current Land Base (The Haudenosaunee have lands in Wisconsin, Oklahoma, and Canada)

Chapter 1

The Problem: Stereotypes

On December 1, 1927, the Grand Council Fire of American Indians made the following presentation to William Hale Thompson, the mayor of Chicago.

> We, therefore, ask you while you are teaching school children about America, first, teach them the truth about the First Americans.... History books teach about Indians as murderers—is it murder to fight in self-defense? ... White men called Indians thieves—and yet we lived in frail skin lodges and needed no locks or iron bars. White men call Indians savages. What is civilization? Its marks are a noble religion and philosophy, original arts, stirring music, rich story and legend. We had these. Then we were not savages, but a civilized race ... The Indian has long been hurt by these unfair books. We ask only that our story be told in fairness.[1]

That was 1927. Have educational materials improved since then? Are the perspectives of culturally rich and diverse American Indian peoples included in textbooks? Do textbooks present American Indians as people, as human beings, with the characteristics of all human beings? Are American Indians included in the history of the nation? The analysis presented in this chapter will cite numerous studies that show that textbooks consistently present standard stereotypes, omissions, and distortions about American Indians

1

rather than multiple perspectives. American Indian perspectives are not evident in textbooks.

This chapter begins with an analysis of the way in which textbooks present American Indians, and chapter 4 provides a culture-based perspective or framework. To begin this analysis we must look at the stereotypes about American Indians in textbooks. Why are these stereotypes so pervasive? What theories support the stereotypes? Why haven't textbooks changed? Why hasn't curriculum that values cultural diversity been developed to replace stereotypes in textbooks? This chapter is divided into three sections. The first section defines stereotypes about American Indians and discusses stereotypes in the media and textbooks. The second section examines stereotypes about the Iroquois in textbooks. The third section includes statements from Cornell undergraduates in the fall of 1991 showing the pervasiveness of the stereotypes to the present day.

Gordon Allport (1958) provided a definition of stereotypes and textbooks in his classic *The Nature of Prejudice:*

> Whether favorable or unfavorable, *a stereotype is an exaggerated belief associated with a category. Its function is to justify (rationalize) our conduct in relation to that category.*[2]

Allport's conclusion from a 1949 study by the American Council on Education states:

> Textbooks used in schools have come in for close scrutiny and criticism. An unusually thorough analysis reports that the treatment given minority groups in over three hundred textbooks reveals that many of them perpetuate negative stereotypes. The fault seems to lie not in any malicious intent, but in the culture-bound traditions which the authors of textbooks unwittingly adopted.[3]

A 1977 study by the Council on Interracial Books for Children, *Stereotypes, Distortions and Omissions in U.S. History Textbooks,* defined stereotypes as "An untruth or oversimplification about the traits and behaviors common to an entire people." Distortions refers to writers who "twist the meaning of history by slanting their presentation of facts . . . and by the omission of information that would alter the viewpoint being presented."[4] Stereotyping occurs "when an entire group is characterized in specific ways and these characteristics are attributed to all individuals who belong to that group."[5]

With these definitions in mind, the next section presents five basic academic stereotypes about American Indians: the Noble Red Man or Noble Savage, the Savage, the Vanishing Race, Living Fossils, and the media image of Generic Indians. Each stereotype will be discussed to provide a definition and indicate the pervasiveness of the stereotype. Studies from 1968–1991 show the pervasiveness of stereotypes across educational materials from K–12 to college level and across subject matters in literature, history, and social studies.

The Noble Savage image is the romantic view of Indians as living close to nature, usually naked, in a simplistic romantic harmony with nature. This stereotype developed in the early days of European invasion and "captured the imaginations of men like Montaigne, Erasmus and Rousseau."[6] In literature, Longfellow's Hiawatha was a combination of the Iroquois "Aywentha," who was one of the founders of the Great Law of Peace, and Ojibwa mythology which he used to create the truly romantic view of American Indians. The romantic image formed the foundation of early views of American Indians, especially in Europe. From the beginning, American Indians were lumped into this generic group and defined as if they were one people. Therefore, the Western Hemisphere was never portrayed to reflect that it was populated by many diverse nations of Native peoples.

Another form of this stereotype is the Indian maiden who proves she was "eager to be of service to the superior white man."[7] The story of Pocahontas illustrates this view and provides the subtle, but pervasive, view that Indians can only better themselves by aligning with or becoming like the European settlers. Harris summarizes the use of these stereotypes in schools:

> Hesitant to change a precedent the early settlers set, the school system combines the best of all choices and mentions a few Indians in history. It praises those Indians who helped white people grab the land or kill other Indians, depicts the rest as blood-thirsty savages who were impediments to progress, and presents the race in general as freaks, now extinct, whose struggle against the oppressors is depicted as a manifestation of the age old struggle of good and evil—the Indians evil, of course.[8]

The savage stereotype depicts American Indians as being inherently, genetically warlike. These are the war-whooping, raiding, scalping savages who attacked the defenseless pioneers. This stereotype does not acknowledge that American Indians had a right to defend their homes and land against invasion by foreigners, as does any nation under attack by foreigners.[9] The people who moved westward across this country did not view the American Indian in a romantic light. To these people, the Indian was a "heathen savage."

> The Puritans with their Christian imperialism contended that God meant the civilized Englishman to win the land from the "heathen" Indian. They theorized that the Indians were descendants of Adam through the Asiatic Tartars who had come to America by a land bridge from northern Asia. Because he had wandered so far, the Indian was far from God and had lost his civilization and law. He was in the power of Satan. The white man came, thus, to the conclusion that the Indian's life as hunter and wanderer was what made him a savage.[10]

The savage stereotype portrayed Indians scalping non-Indians, (a practice that was introduced to New England Indians by the Puritans in 1637), as thieves, and drunkards. This view of Indians as killers was used to "justify the slaughter," the genocide, of American Indians.[11]

In the savage stereotype, American Indians are presented as an obstacle to "progress," that is, the inevitable settlement of this land by Europeans, and as being quite incapable of becoming civilized. The theme of the "brave colonists surrounded by dangerous savages" is prevalent in textbooks.[12]

Virgil Vogel's study of 100 history books defined the theme of disembodiment, which portrayed the American Indians as subhuman, nomadic, an obstacle to civilization, as wild animals or savage men. The theme of Indians as primitive hunters who did not "develop the land" led to the theory that the land was unoccupied, except for roaming hunters. Vogel defines defamation as the focus on so-called inferiority traits such as being, lazy, filthy, and capable of extreme cruelty.[13] The savage image is the most enduring stereotype of American Indians. It pervades literature, history, social studies, media, cartoons, even children's toys and contemporary movies.

During the late 1800s the general policy was to either eliminate or assimilate American Indians. Once again, the romantic notion emerged as the view of American Indians as a vanishing race became popular. This stereotype flourished from the late 1800s into the early 1900s when it was believed that American Indians were becoming extinct. Ethnographers engaged in a hectic flurry of activity during this time period in an effort to document as much as possible of American Indian life before it vanished. Edward C.

Curtis travelled the country carrying along "Indian" clothes and posing Indians in them for his famous photographs of "real Indians" who were soon to vanish. He created an image. Helen Harris's study, "On the Failure of Indian Education," linked the vanishing race stereotype to James Fenimore Cooper's version of Indians as a people who would inevitably become extinct. Harris cites Emerson, Longfellow, and Thoreau as contributing to this view.

The living fossils stereotype perpetuates the idea that any surviving American Indians are mere remnants of a once proud people. This view keeps American Indians frozen in the past (that is, the pre-1890 images by artists such as George Catlin) as the dominant visual image of American Indians.[14] The fact that the Museum of Natural History in New York City focuses on dinosaurs and Indians certainly assaults the senses of contemporary American Indian people. The implicit message is that just as dinosaurs are extinct so are American Indians. This image is the standard American visual image of "Indians" in commercials, movies, toys, and books. The end result of the living fossils stereotype denies the diversity and dynamic nature of Native American cultures. Textbooks reinforce this stereotype, which leaves children without enough information to process the reality that American Indians are alive and well today, living contemporary lives, and maintaining their complex cultural heritage.

> If we were to ask most teenagers and many adults what an Indian is, we could predict the answers. He was an early inhabitant of America who rode horseback, hunted buffalo, wore a feathered headdress and beaded buckskin and lived in a teepee. This is the picture of the Indian which Buffalo Bill and his original Wild West Show made famous . . . and which has been kept alive by television, museum displays.[15]

The generic stoic Indian is portrayed as a silent, humorless, granite-faced cigar store Indian. This is the Indian who says "How" and "Ugh." A generation of Americans grew up watching the Lone Ranger and Tonto, which solidified this stereotype in their minds. A League of Women Voters study of kindergarten and fifth grade students found seventy-six percent of kindergarten students had already learned the generic image of the Indian in feathers and teepees.[16]

The generic stereotype was already evident by the 1850s. *The American Child's Pictorial History of the United States* (1860), which was adopted as a textbook, depicted "the Indians of New England, Virginia, and Roanoke Island as living in tipis and wearing flowing-feather bonnets of Plains Indian type."[17]

"I is for Indian" is a standard identification in alphabet books and classrooms. The "Indian" portrayed is often a caricature or an animal dressed up like an Indian. This use of "I for Indian" dehumanizes American Indian peoples by placing them into a category of objects. These inventions or creations of an image of American Indians and the perpetuation of these images continue to this day. These stereotypes lump all Native Americans into one mold, thus denying the immense diversity of American Indian Peoples.

Media

During the 1800s, dime novels, literature, art, and Buffalo Bill's Wild West Show combined to create the image of American Indians as savages. This image has continued "from the era of Columbus up to the present without substantial modification or variation."[18] Saturday morning cartoons include many war-whooping, tomahawk-wielding, painted Indians on the warpath. Cartoons and picture books with "I" for Indian are young children's earliest exposure to stereotypes about American Indians.

The stereotypical image continues to pervade the movie industry. *Dances With Wolves,* the recent blockbuster movie on American Indians is, of course, situated in the late 1800s, and although the Lakota people are presented as human beings, they are doomed to extinction. The basic theme of American Indians as people of the past continues. Although the current trend in the movies tends to portray American Indians as real human beings and their societies as making sense, that image is always mitigated by the appearance of another tribe of Indians who uphold the savage stereotype. Thus, "a few" Indians are good and the rest are bad. This so-called advancement is, in fact, a repeat of the Pocohantas theme that some Indians are good and can be converted to civilization, but the other Indians, the savages, are incapable of civilization.

How pervasive are these stereotypes and where do most people receive their information about American Indians? A 1991 study by Rouse and Hanson, *American Indian Stereotyping, Resource Competition, and Status-based Prejudice,* provides an examination of stereotypes and prejudice in college students at university settings in Texas, North Dakota, and Wisconsin. The study states that ignorance-based prejudice can be "modified by new information presented by a legitimate source (e.g., teachers)."

Stereotypic cultural beliefs about all Indians living in tipis, being warlike, migratory hunters, carrying tomahawks, carving totem poles, and speaking "Indian" are modified when students are presented with more accurate information about Indian history and ethnography. Likewise, students will accept an instructor's or text's authority.[19]

The authors cite mass media and advertising as the major sources that perpetuate stereotypes about American Indians.

They were looking at status-based prejudice that exists where communities are competing for resources such as hunting, fishing, or water rights. Receiving new information does not change this type of prejudice as people will discount the new information as being biased towards Indian peoples. Questionnaires were given to students taking introductory courses in sociology and anthropology. The questionnaire assessed three areas: 1) concepts held about American Indians; 2) students sources of information about American Indians; and 3) an opinion and knowledge survey to measure factual knowledge, general orientation, and victim-blaming.

> They tended to share some traditional cultural stereotypes that reflect a pervasive generic "folk ethnography" promulgated by the media and recreation and leisure industry. Correspondingly, for all three samples, *the* highest ranked source of information about *American Indians was TV/movies.*[20] (emphasis added)

The study found that negative stereotyping varied according to the presence of Native Americans in the area and the level of competition over resources.

The Rouse and Hanson study indicates a uniformity in the ignorance about American Indians, because young people are greatly influenced by the media, and that making curriculum change in areas closer to Indian reservations will be more difficult than in areas where there are fewer American Indians. The state of Wisconsin passed a law in 1989, which was implemented in September 1992, requiring that units about Wisconsin Indians be taught at least two times in elementary grades and once in the high school. This law was enacted to deal with the public ignorance of treaty rights apparent in the conflict over fishing rights in northern Wisconsin.

Hirschfelder declares "Non-Indian writers have created an image of American Indians that is almost sheer fantasy." The media perpetuate these images and keep them firmly in the public domain through movies, television shows, commercials, and cartoons.

Textbooks

Were these stereotypical images of Indians wandering in the wild forest limited to academia? No, they were, and continue to be, perpetuated in the textbooks of American school children.

> What American authors preached in their novels, plays, and poems about the inevitability of civilization superseding savagery, regardless of nobility, American school children learned in their textbooks.[21]

Ruth Elson conducted a study of more than a thousand textbooks used in American schools during the nineteenth century. Her study, *Guardians of Tradition, American Schoolbooks of the Nineteenth Century,* reveals the early textbook depiction of American Indians. Common themes included the romantic dignified noble savage and the cruel warlike savage. The theme of the American Indian as a nomad or wanderer in the wild is presented in an 1895 Reader which

> describes the American forest at the time of the American Revolution: "where wild beasts and scarcely less savage Indians roamed in their freedom."

Along with the descriptions of these barbaric people were

> profuse illustrations of their cruelty in highly dramatic tales of Indian warfare . . . many books present pictures showing an Indian about to tomahawk a woman with a child in her arms."

In fact, Elson states that "every Reader includes at least one detailed account of gory Indian warfare."[22]

The justifications and/or rationalizations for the conquest of the Indians outlined in these textbooks included: 1) God's natural law, manifest destiny, the march of civilization; 2) the inherent inferiority of Indians and superiority of Europeans; 3) early cruelty to Indians as part of British colonialism; 4) Spanish treatment of Indians as more cruel than the English; 5) the implication that Indians were incapable of civilization; and 6) the inevitable extinction of Indians.

An example of the inevitable march of civilization can be seen in the way the removal of the Cherokees has been handled in textbooks. The Cherokees' battle to resist removal from their lands was portrayed as refusing civilization, as refusing progress, even though the Cherokees had become quite wealthy farmers. That certainly was a one-sided presentation of a significant historical event. The impact of such textbooks on school children in the nineteenth century, indeed up to contemporary times, results in students never questioning colonial actions taken against American Indian Nations. The analysis of nineteenth-century textbooks illustrates the origins and consistency of the major themes in textbooks in the twentieth century. Textbooks appear to be, even given great spans of time, virtually unchangeable.

The review of stereotypes in textbooks in this chapter provides documentation showing that stereotypes have not changed substantially in textbooks. The worst words, such as *savage* and *terror*, may have been omitted recently, but the basic image of savages, as Indians destined to be overrun by civilization, and of Indians as belonging to the past, continues in textbooks.

Henry and Costo document another instance of American Indian objections to stereotypes in educational materials. On August 19, 1965, a group of American Indian scholars and historians, calling themselves the American Indian Historical Society, addressed the California State Curriculum Commission to make a statement on California textbooks.

> We have studied many textbooks now in use, as well as those being submitted today. Our examination disclosed that not one book is free from error as to the role of the Indian in state and national history. We believe everyone has the right to his opinion. A person also has the right to be wrong. But a textbook has no right to be wrong, or to lie, hide the truth, or falsify history, or insult and malign a whole race of people. That is what these textbooks do. At best, these books are extremely superficial in their treatment of the American Indian, over-simplifying and generalizing the explanation of our culture and history, to the extent where the physical outlines of the Indians as a human being are lost. Misinformation, misinterpretation, and misconception—all are found in most of the textbooks. A true picture of the American Indian is entirely lacking.

Henry documents testimony on January 4, 1969, before the Senate Committee on Indian Education:

> There is not one Indian in this country who does not cringe in anguish and frustration because of these textbooks. There is

not one Indian child who has not come home in shame and tears after one of those sessions in which he is taught that his people were dirty, animal-like, something less than human beings. We Indians are not just one more complaining minority. We are the proud and only true Natives of this land.[23]

In 1970, Henry and Costo's *Textbooks and the American Indian* documented the work of thirty-two American Indian scholars evaluating more than 300 books used in American schools. The results of this study are succinctly stated: "Not one could be approved as a dependable source of knowledge about the history and culture of the Indian people in America."

In 1977, the Council on Interracial Books for Children, funded by the Carnegie Corporation, conducted a study of thirteen U.S. history textbooks published between 1970–75. This study examined the texts for racism, sexism, stereotypes, distortions, and omissions. Native Americans were included in this study of African Americans, Asian Americans, Chicanos, Puerto Ricans, and women. For each group, the text provides an explanation of terminology used and a brief historical background.

The section on Native Americans includes twenty-six content areas that were examined and provides quotes from specific texts, a commentary on each quote, and references. The section also provides a "Native American Textbook Checklist" designed for teachers to use in evaluating texts. In the section on "The 'Indian Image'" the authors provide a concise statement on U.S. history textbooks:

Native Americans . . . have always been visible in history textbooks—or at least an objectified image of "Indians" has been visible. They were there, first to be "discovered" by Columbus, then to "lurk" in the "wilderness," "attack" wagon trains, "scalp" pioneers, and finally—with the buffalo—to "vanish from the scene." Granted, textbooks provided a few "friendly Indians" to offer food or guide services at critical moments, but these were the contrast to the "savages" who hindered, but never halted, the inexorable tide of Euro American "progress."[24]

In 1978, Jesus Garcia conducted a longitudinal study of eighth-grade U.S. history texts in California (1956–76), analyzing stereotypes of American Indians in textbooks. The study examined 1,900 textbook statements and found that stereotypes had not changed significantly between 1956–76.

In a 1984 study of "Native Americans in Elementary School Social Studies Textbooks" which examined thirty-four textbooks used in grades K–7 in Virginia, Ferguson and Fleming state:

> One important element of schooling that is involved in the shaping of attitudes of children is the textbook. Teachers without specialized training in a subject content area rely heavily on textbooks as a source of information. If the textbooks are inaccurate or biased, this misinformation or bias will likely be transmitted to students.[25]

They used a three-part evaluation based on ten key concepts, evaluative terms, and a picture analysis. Their conclusions on the ten key concepts revealed that although content differed, key concepts such as differing perspectives on land ownership were ignored. They also found that little attention was paid to contemporary American Indians and current-day issues. Use of less biased language, by eliminating such words as "savage," "fierce," and "terror," was cited as an improvement; however, they reached the conclusion that textbooks need much more improvement.

Another study, *The Indian Versus the Textbooks: Is There Any Way Out?* by Frederick Hoxie of the Newberry Library (1984), examined thirteen U.S. history textbooks at the college level. "I discovered that despite the changes and improvements of the past fifteen years, many of the distortions and inaccuracies referred to by Vogel . . . persist."[26]

> My sample indicates that textbook authors simply ignore new information. This is particularly true when that information threatens cherished preconceptions about the American past. It is easier to add a brief biography of Geronimo to a chapter on the West than to surrender our self-image as tamers of the wilderness or settlers in a virgin land. . . . The appeal of the "lonely settler in the howling wilderness" motif is made plain by a quick check of textbook descriptions of Plymouth Colony. Only three of the thirteen texts identify Squanto as the individual who stepped forward to save the colony from starvation in the spring of 1621. And of those three, only one tells us that Squanto had previously been captured and taken to England. . . . With that information— and the knowledge that Squanto returned from England in 1619 to find his village wiped out by an epidemic—we get a much fuller picture of the man and his motives.[26]

Hoxie provides four recommendations that would enable writers to present American Indian peoples as "coherent, multi-faceted actors in American history." The first point acknowledges that Native American cultures are "nonwestern," and are based on communal cultural values. He states that writers must understand that American Indian cultures are "rooted in the obligations of kinship," with traditional values and ceremonies that structure their way of life. The second point recommends presenting encounters between American Indians and Europeans as "cumulative interaction." Third, he recommends:

> [O]rganizers of textbook projects need to purge their books of inaccurate and misleading shorthand references that suggest Indians lack coherent motives for their actions. Indians did not "wander" or "roam." Tribes did not live in isolation. Many Indians experienced military defeat, and all Indians witnessed changes in their cultural life; these facts do not mean that Indian cultures were necessarily destroyed or corrupted through contact with Europeans.

And, fourth, he recommends utilizing research from cultural anthropology that would help *"to understand and explain the coherency in all cultures."* (emphasis added)

Hoxie identifies a major problem in textbooks as their treatment of what he calls Presence and Absence. He defines the pattern or typical organization of textbooks with regard to American Indians:

> Indians appear at the time of discovery, in skirmishes accompanying early settlement, in the revolutionary war (as British allies), in descriptions of the Old Northwest and the War of 1812, during removal, at the Little Big Horn and Wounded Knee, as beneficiaries of the Indian New Deal, and as militants at Wounded Knee II and Alcatraz.

Actually, several of these events are usually lacking in most textbook coverage. The twentieth century is generally not included. "For the most part Indians simply cease to exist in texts following the battle at Wounded Knee." Hoxie refers to the absence or lack of information on contemporary American Indians as "historical selectivity."[27]

In his section on Indian Legacy, Hoxie discusses at length the plurality of American culture. "Texts have difficulty admitting that American history is not the story of one group. U.S. history is the story of many groups who met and affected one another in the North American environment." He states that "defining a workable and honest view of the Indians' role in the development of American society and culture is the key to integrating Indian materials into courses and texts on the history of the United States."[28]

G. Patrick O'Neill (1987) reviewed ten studies of stereotyping of American Indians in history and social studies textbooks published between 1976–84 and concluded that textbooks had not substantially improved in the last twenty years.

In his 1989 study of fifteen high school literature texts adopted in South Carolina, James Charles evaluated these texts to find out if they reinforced stereotypes about American Indians. The four categories of stereotypes found in this study were: Noble Savages, Savage-Savages, Generic Indians, and Living Fossils. His initial analysis showed American Indians as well represented in the texts examined. However, Charles conducted what he calls a "deeper analysis" and found a "lack of proportional representation" of the following: 1) the contemporary genres (written) of poetry, drama, and essay; 2) traditional (oral) genres of song-poems and oral narratives; 3) regional affiliations of authors; and 4) form and content aspects of literature at particular grade levels. An interesting point in his analysis is that song-poems are often brief, because, citing a Papago explanation, "the song is so short because we know so much." This should "remind us that in oral poetry, more than ever, we need to know the cultural background, the complex allusions, and the numerous associations evoked in the native audience or participant." He recommends including more oral narratives and oratory, including contemporary examples.[29]

A Study on the Iroquois in Textbooks (1975)

Only one article deals specifically with the Iroquois, *The Treatment of Iroquois Indians in Selected American History Textbooks,* by Arlene Hirschfelder. In this study Hirschfelder examined twenty-seven history textbooks published in the 1950s and 1960s, and found that three entirely omitted the Iroquois, and three mentioned them only once. Only one writer mentioned the six distinct nations of the Iroquois, but the information was in a picture caption, not in the main body of the text. Including information on diverse peoples in America within a picture caption or box presents them as peripheral to American history. Inclusion in the main body of the text presents them as a significant part of history.

Hirschfelder examined various themes in her article showing how these themes were omitted or distorted in the texts. Under the theme identified as behavior, the Iroquois were presented as "warlike," instilling "terror," as "barbarous nations" with "ferocious vitality," who were "constantly on the warpath." Hirschfelder supports the claim that the Iroquois tortured those people who were not adopted into the Iroquois nations, but she also shows this one-sided statement as

> deceptive because torturing one's enemies was an accepted code of behavior among Indians and non-Indians. A misimpression that Iroquois were particularly brutal to enemies has been created by the omission of pertinent data regarding the similar behavior of non-Indians.[30]

Certainly, presenting information on "warlike" behavior would require a worldwide study that would show that in any culture, in any war, in any time period, "warlike" behavior contains wartime atrocities.

Hirschfelder notes that in addition to being warriors, Iroquois men were also hunters, craftsmen, physicians, politicians, dancers, and religious leaders. This writer would add to the list the men's vital roles as orators, singers, storytellers, fathers, uncles, and grandfathers in Iroquois families.

The women's roles are often neglected in textbooks, and only one picture caption mentioned women storing and preparing food. Hirschfelder provides information on the role of Iroquois women:

> Iroquois women were the farmers in their culture as well as clothesmakers, dispersers of herbal remedies, cooks, and dancers in ceremonials. They also had an important political role because of their power to name and remove Confederacy Chiefs.

She quotes Hazel Hertzberg's comparison of Iroquois women and European women of the same time period:

> There is no question but that at the time of European contact, the Iroquois woman occupied a higher, freer, and more influential place in her society than did the European woman in hers . . . the Iroquois woman's position was securely based on her leadership in the family and in agriculture.[31]

Hirschfelder discusses settlement patterns that focus on the longhouse structure. She mentions that students are not taught the cultural basis for living in the longhouse and are left with idea that the Iroquois continue to live in longhouses.

The differing views of land ownership and land usage between Iroquois and Europeans were not discussed except to say Indians were ignorant of European land values, and private property, thus once again failing to present Iroquois cultural values of communal land usage.

Only three authors provide some explanation of the origins of the Iroquois Confederacy. The debate over exactly when the league was founded is mentioned in only one text. Some have tried to pinpoint a date and their guesses have been quoted over and over as fact, but in truth, the Confederacy originated long before Europeans arrived and a date cannot be established. Hirschfelder states: "Moreover, no authority has determined the precise time the Confederacy was formed."

Only one writer refers to the influence of the Iroquois Confederacy on the founding fathers. Twelve of the textbooks mention the Iroquois attended the Albany Conference of 1754, but none of these discusses the vital role of the Iroquois in that crucial meeting. The purpose of this conference was to improve relations with the Iroquois, thereby strengthening colonial position. Benjamin Franklin was influenced by the Iroquois Confederacy as a form of government. Hirschfelder found that not one textbook refers to this cultural interaction. She quotes the now famous statement of Benjamin Franklin regarding the Albany Plan of Union:

> It would be a very strange thing, if six nations of ignorant Savages should be capable of forming a Scheme for such a Union, and be able to execute it in such a Manner, as that it has subsisted ages, and appears indissoluble; and yet that a like Union should be impractical for ten or a Dozen English colonies, to whom it is more necessary, and must be more advantageous; and who cannot be supposed to want an equal understanding of their ignorance.[32]

Franklin's statement is clear in presenting his case and his knowledge of the Iroquois Confederacy as he admonishes the colonies to unite.

Hirschfelder found only two textbooks contained references to current-day Iroquois people and these references did not provide a map of current-day reservations.

> It is not surprising that many Americans, both young and old, are not aware that Iroquois reservations exist in upstate and

western New York because a basic source of information, American history textbooks, has omitted this information.[33]

Hirschfelder concludes that textbook "information is inaccurate, ethnocentric, misleading, insufficient, or altogether missing from the narrative."[34]

The 1990s: The Stereotypes Continue

In the 1990s one would expect that there have been great improvements in textbooks and that the average American has by now received a better education about American Indians. Have all these stereotypes, omissions, misinformation, and inaccuracies been corrected? One would hope so. However, in a seminar taught in the fall of 1991 at Cornell University using Jack Weatherford's *Indian Givers,* students made the following statements in their final papers:

> My previous knowledge of Indians, received through my elementary and high school education, was sketchy and often biased. . . . The only time students were given the opportunity to grasp that Indians had actually accomplished something in the Americas was when we were given the chance to try (with paper bags and popsicle sticks) to recreate their art and housing.
>
> As is often the case with United States society, it is our tendency to assume that Americans and Europeans were the source of the American culture. Having grown up in a white, middle class society, I too must admit to some ignorance of this sort . . . I feel that the more important point of this book is that we, as American citizens, have never been told this information before. We are told of scalping, teepees, war dances, face paint, "ugh" and "how," but rarely do we learn that so much of modern American culture owes its origination to the Native Americans. We only hear that the Indians got in the way of our development and settlement of the American continents, rarely are we told of how much they contributed and helped us in the construction of our culture.
>
> In elementary school, I remember that we would only discuss the Indians around Thanksgiving and even then, it was arts and crafts. We would make headdresses out of construction paper. We didn't really understand Indian culture, we just were taught that they know how to grow corn and other crops.

This country's "educational" system has misled us to believe that before Christopher Columbus brought"civilization" to America that this was a land of barbaric savages.... As I read this I became angry with myself having discovered just how vulnerable I had been to our educational system which had successfully manipulated me into believing that the English Settlers who left a monarchial society had a concept of what democracy actually was without the aid of any other peoples. To have recently learned that the Indians had achieved the highest cultural development of liberty, freedom and individuality, while having had been lied to all these years about their savage-like nature reveals the first true civilized nation. After having been enlightened, I am afraid that I can never fully accept what an instructor deems to be truth without my own ample research.

My image of the American Indian used to be two dimensional, and I am ashamed that to me Indians were no more real than their cardboard pictures at Thanksgiving time.

These statements show the pervasiveness of stereotypes in textbooks. These young college students, products of the American education system, are protesting the same stereotypes the elders identified in 1927. From my own experience teaching on the college level and discussions with my colleagues across this country, we continue to hear the same statements as those above made by our students.

No matter what grade level—kindergarten to college level—whether in history, literature, or social studies, the stereotypes, omissions, and distortions about American Indians continue to pervade educational materials. What is the basis, the underlying assumption behind these images of the noble savage, savage savage, or the vanishing race? Why have American textbooks presented such a distorted image of American Indians? Why do these images consistently deny the cultural complexity, the diversity, of American Indian peoples? Why do the texts provide perspectives not based in reality, but instead provide "invented" views and images of American Indians? The next chapter delves into the reasoning behind the development and persistence of these stereotypes.

Chapter 2

Cultural Evolution:
The Theories behind the Stereotypes

The invented stereotypes of American Indians generally found in textbooks and the media diametrically oppose the actual reality of American Indian peoples who live culturally rich, complex, and diverse lives. Where, and why, did the stereotypical images of American Indians develop? What theoretical basis has been perpetuated to support, rationalize, and justify these false images?

First, the Europeans developed their own theory on the origins of American Indians as wanderers across the Bering Strait, which denied any knowledge of, or respect for, the cosmology of the indigenous peoples of this hemisphere. Second, the myth that this country was a vast, empty, raging wilderness occupied by only a few savage hunters denied the existence of approximately 500 separate and unique indigenous peoples, many of whom engaged in complex agricultural societies.

These two theories present an image of American Indians that is so far removed from the reality as to be unrecognizable by American Indian people. Yet, it is the image that the average American recognizes and believes.

First, we will examine the Puritan's theory surrounding the origins of American Indians.

> Almost universally it was agreed that the Indians were of the race of men, descendants, in order, of Adam, Noah, and those Asiatic Tartars who had come to America by a land-bridge

from northern Asia. This opinion, orthodox in the seventeenth century, allowed Puritans to account simply for the savage, heathenish state of the Indian. Was he not perhaps the fartherest of all God's human creatures from God himself? Descended from wanderers, had he not lost his sense of civilization and law and order? Had he not lost, except for a dim recollection, God Himself? And wasn't he, as a direct result of this loss, in the power of Satan?. . . The Puritan writer on the Indian was therefore less interested in the Indian's culture than in the fallen spiritual condition which that culture manifested.[1]

Thus, the distortions began with a religious definition of American Indians. This theory did not allow room for American Indian beliefs, for their cosmology, because they were labelled as subhuman wanderers from the very first contact with the Pilgrims. This religious definition extended to land rights:

> For here the Puritans carried to its extreme the logic of seventeenth-century Christian imperialism. God had meant the savage Indians' land for the civilized English and, moreover, had meant the savage state itself as a sign of Satan's power and savage warfare as a sign of earthly struggle and sin. . . . Convinced thus of his divine right to Indian lands, the Puritan discovered in the Indians themselves evidence of a Satanic opposition to the very principle of divinity.[2]

Did God approve of this logic? John Winthrop, in 1634, clearly justified God's giving the land to the invading Europeans by implying that the smallpox epidemics that killed 50–95 percent of the native peoples were God's will.[3] Did this religious definition change over time? Thomas Jefferson, writing in 1784 in *Notes on the State of Virginia,* described American Indians as savages who were of Asiatic origin, except the Eskimos who were believed to be of Scandinavian stock. In 1816, this standard view continued the idea that the native populations of this hemisphere were part of the lost tribes of Israel.[4] Thus, the ancient cosmology beliefs of the diverse American Indian peoples of this hemisphere were not even acknowledged, much less incorporated into American textbooks.

The land bridge theory was first proposed "by Jose de Acosta in 1590 in his *Natural and Moral History of the Indies.*" The migration theory was based on biblical beliefs and the diversity of American Indians was attributed to "the post-Flood migrations of Noah's three sons and their progeny to various parts of the globe."[5] The Tower of Babel story was used to explain language differences.

Why was it necessary to develop a theory explaining why American Indians were different? Because, as Berkhofer states, the theory was one of monogenesis, that is, *all* humans were created according to biblical beliefs. The idea that the peoples of this world were the result of polygenetic origins, many separate creations, was not acceptable.

Today, standard textbooks present the Bering Straits theory as fact and do not acknowledge any other perspective on the origins of American Indians. The Puritan "theory" has thus become fact in textbooks. Archeological researchers have spared no expense in attempting to provide scientific evidence to support the land bridge theory.

The second view of American Indians as wanderers in the vast, unoccupied wilderness developed as the religious view was expanded into an economic viewpoint. The theory behind the cultural evolution scales marking mankind's progress from the lower primitive, savage, barbaric state to the higher state of civilization had been well accepted in the early contact era. This scale was imposed on all American Indians when they were labelled savages. The Europeans applied this scale to American Indians in order to describe what they deemed mankind's universal progress from a heathen, or pagan, early state characterized by hunting to the highly valued agrarian lifestyle. In fact, they saw American Indians as reflections of themselves in an early evolutionary state.[6]

Two problems, according to European standards, held the Indian back from obtaining higher levels of "progress": they were mere hunters lacking an agrarian lifestyle, and they had no notion of private property. To understand these two views it is necessary to look at Lewis Henry Morgan's economic and technological scale of cultural evolution.

Morgan's Evolutionary Theory (1818–81)

Civilization:	phonetic alphabet/writing
Upper Barbarism:	iron smelting, use of iron tools
Middle Barbarism:	domesticated animals (Old World), corn with irrigation, adobe, stone brick buildings
Lower Barbarism:	pottery
Upper Savagery:	bow and arrow hunting
Middle Savagery:	fishing, use of fire
Lower Savagery:	speech, subsistence on fruits and nuts

—Malefijut, *Images of Man,
A History of Anthropological Thought*: 150.

Morgan's scale was not new, it only codified the invented view of mankind's progress through time from lower levels to upper levels based on economic developments. This scale, and others, were invented to reinforce the European conception of "progress" to civilization, and to place peoples of the western hemisphere on the bottom rung of the scale.

Many observations by early explorers provided evidence that American Indian people were agriculturalists. Why then has the stereotype of savage wanderers survived in the face of evidence to the contrary?

> The basis of their understanding had long been part of the grand rationale of westward-moving colonialism. This was the tradition of the natural and divine superiority of a farming to a hunting culture. Universally Americans could see the Indian only as hunter. That his culture, at least the culture of the eastern Indians whom they knew best until the second quarter of the nineteenth century, was as much agrarian as hunting, they simply could not see. They forgot too, if they had ever known, that many of their own farming methods had been taken over directly from the Indians whom they were pushing westward. One can say only that their intellectual and cultural traditions, their idea of order, so informed their thoughts and their actions that they could see and conceive of nothing but the Indian who hunted. Biblical injunction framed their belief; and on the frontier practical conditions supported it. The Indian with his known hunting ways needed many square miles on which to live, whereas the white farmer needed only a few acres. The latter way was obviously more economical and intelligent; it was essentially the civilized way. Therefore the Indian would have to move on to make way for a better and higher life.[7]

The hunting or nomadic theory was applied to the Iroquois in the early colonial days. Schoolcraft in 1846 continued the image of the Iroquois as just hunters, despite the evidence that in 1779 Congress had instructed George Washington to destroy the Iroquois through the destruction of Iroquois crops. The journals of the Sullivan-Clinton campaign document vast fields of corn, beans, and fruit orchards of up to 1,500 trees.[8] Even though the early historians were confronted with the reality that American Indians were hunters, gatherers, and fishermen and had developed a highly sophisticated agricultural society, they continued to perpetuate the hunter image.

The Haudenosaunee engaged in a seasonal and cultural cycle of hunting, fishing, gathering, and agriculture, the combination of which did not fit neatly into the categories on Morgan's scale. Intricate and sophisticated scientific knowledge of nature's seasonal cycles was required to engage in this lifestyle. One had to know the correct time to gather certain plants for medicines and the proper use of these plants as medicine, which is a lifetime study based on apprenticeship with a knowledgeable elder. The proper time to gather and use plants and berries for food required learning the seasonal cycles. Knowledge of the moon's cycles was essential to planting, as was the science of nutrition for the earth and people. The agricultural lifestyle required complex technological skills and tools, experimentation in cross-fertilization of corn to develop varieties, and methods of food preservation. Fishing required technological skill as well as knowing when specific fish were abundant. Hunting involves an interdependent relationship with the woods and animals. Yet, even with all this information on the Haudenosaunee way of life, they were labelled as mere hunters.

The Jeffersonian agrarian policy was the model ingrained in the American mind as the way to true civilization. He saw agriculture, or the agrarian state, in a different light than did the Haudenosaunee. To Jefferson agriculture meant farming, the components of which were plows, fences, and domesticated animals, which required less land and, according to him, more productive use of the land, than the wild hunter state. Jefferson perpetuated the evolutionary scales in the early founding of this country because acknowledgment of the agricultural societies of American Indians would have counteracted his beliefs.

> In his census of the New York Iroquois communities published in 1847, Henry Schoolcraft described a livelihood that mixed a diversity of activities: the cultivation of various crops, hunting and gathering, the raising of livestock, and harvest work on neighboring white farms. This and other evidence, nevertheless, did not stop American intellectuals and politicians from continuing to portray Indian economies as fundamentally in a hunter state. In fact the increasing drive for more Indian lands gave impetus to the use of Jeffersonian theory, both friends and foes of American Indians invoking the image of nomadic hunters.[9]

The increasing drive for more land underlies the perpetuation of the hunter image, and Jeffersonian agrarian theory supported the cultural evolution scale. Thus, the invented image of American

Indians was used to justify taking of so-called "under-used" lands which were put to better use by agrarian settlers. It was a clash of two extremely different world views. American Indians had been defined in purely Western ideological terms, which suited the purposes of the Europeans to acquire the land.

The world view of Europeans included a fear of nature, which was translated into the drive to conquer nature. "This fear of the 'wilderness' has been traced by Frederick Turner III far back to its Judeo-Christian roots."[10]

During the debates of the early 1800s on removal of the Cherokees from Georgia to west of the Mississippi, and on treaty making with American Indian nations, Governor George Gilmer of Georgia stated:

> Treaties were expedients by which ignorant, intractable, and savage people were induced without bloodshed to yield up what civilized people had the right to possess by virtue of that command of the Creator delivered to man upon his formation—be fruitful, multiply, and replenish the earth, and subdue it.[11]

Kirkpatrick Sale describes "Europe's fear of most of the elements of the natural world," which "estranged" them from the "natural environment and had for several thousand years" and resulted in "depleting and destroying the land and waters it depended on."[12] This fear and rejection of nature, this "separation from the natural world, this estrangement from the realm of the wild, I think, exists in no other complex culture on earth."[13] Yet, there was a time when Europeans believed that "gods and spirits inhabited the elements of nature—trees, certainly, streams and rivers, forests, rocks—that nature was sacred because God was immanent in all that He Created."[14] The time when European people held this view of nature was in the long-distant past, when they were "pagans," which was so far in their past that there was little, if any, surviving recognition of the value of nature.

The results of the view of mastery over nature, of fear of nature, of dominance and exploitation in Europe were "deforestation, erosion, siltation, exhaustion, pollution, extermination, cruelty, destruction, and despoliation."[15] In 1492, Columbus brought these views of wild men and the fear of nature with him to this hemisphere.

> To have regarded the wild as *sacred,* as do many other cultures around the world, would have been almost inconceivable

in medieval Europe—and, if conceived, as some of those so-called witches found out, certainly heretical and punishable by the Inquisition.[16]

Given this dramatically different view of nature it is not difficult to understand the clash of world views that took place upon the encounter of the two civilizations in 1492 and thereafter.

Adam Kuper's critique of anthropology, *The Invention of Primitive Society, Transformations of an Illusion,* analyzes the anthropological idea of primitive society that developed in the 1860s and 1870s. He reviews the idea of primitive society which originated as a "fantasy which had been constructed by speculative lawyers in the late nineteenth century."[17] Kuper states that the orthodox view is no longer accepted and that the "modern view is that there never was such a thing as 'primitive society.'" In fact, he says, "it does seem likely that early human societies were indeed rather diverse."[18] The purpose of his book was to "remove the constitution of primitive society from the agenda of anthropology and political theory once and for all."[19] The current orthodoxy tends to support relativism to "avoid culture-bound misapprehensions, to achieve phenomenological validity."[20]

Let's hope the move to relativism, although it will be excruciatingly slow, will allow textbook writers to utilize the ethnographical information that was gathered by early anthropologists to provide insight and documentation to a more realistic picture of American Indians. Historical documents contain the evidence to prove American Indians were not nomads wandering in the vast wild unoccupied forest, but indeed, had settled, diverse societies with an economic base of agriculture, hunting, fishing, and gathering. The evolutionary scale is simply a fantasy, as Kuper reveals.

Chapter 3

The Theories Become
the Standard Curriculum

Defining Culture and Cultural Relativism

Where do we begin the process of developing a different perspective
on how to study another culture in a respectful manner? In order
to have another perspective we must dispense with evolutionary
scales, invented images, and stereotypes by engaging in theoretical
framework that values diverse cultures.

First, indigenous cultures are defined as those Indigenous
Peoples who continue to maintain their own cosmology, world view,
language, ceremonies, government, economic system, health systems, and traditions which are rooted in their specific land base
and which have existed from antiquity. Indigenous peoples have a
specific way in which they perceive reality, and that reality is based
in ancient beliefs about how this world originated and how human
beings should conduct themselves on this earth. Indigenous cultures have deeply rooted complex sets of beliefs, customs, and traditions that are culturally specific and those beliefs are maintained
to the present day because those beliefs are integral to life as the
people of the culture understand life to be.

Until recently, the standard definition of culture was singular,
and only Western cultures qualified as civilized culture. Goodenough
provides a concise discussion of culture as the higher end of the
scale from savagery to civilization. "The degree to which people

differed in their customs, beliefs, and arts from sophisticated Europeans was a measure of how ignorant and uncivilized they were."[1] He explains that within this cultural hierarchy was an "arrogantly ethnocentric theory developed by the imperial elites of the nineteenth century," including Lewis Henry Morgan.[2] Stocking's definition broadened the concept of culture to include:

> characteristic manifestations of human creativity: art, science, knowledge, refinement—those things that freed man from control by nature, by environment, by reflex, by instinct, by habit, or by custom.[3]

Culture, as defined by Western societies, involved control over nature and valued those items created by human beings, which contrasts with indigenous peoples' definition of culture which values nature and understands that human beings are interdependent with nature. Within anthropology, Boas began to broaden perspectives by moving away from the singular evolutionary view of culture to recognizing the diversity of cultures.[4]

The theoretical background for cultural relativism, that is valuing diversity, is not new, it just has not become widely accepted or instituted. Boas was one the first people to question the evolutionary view as he started the process to change the evolutionary scale by promoting and valuing diversity.

> For Boas, the anthropological perspective centered on the concept of cultural relativism, *where cultural patterns should be interpreted within the context of the societies in which they occur rather than through the ethnocentric lens of Western culture.*[5] (emphasis added)

Boas was the founder of what is called the "American" school of anthropology in the early 1900s. At Columbia University, he introduced this new way of thinking about culture within anthropology, and by World War I he had established a new perspective on culture:

> Soon other professional anthropologists in the United States followed Boas and his students in repudiating raciology and evolutionism and espousing the idea of culture as the way of understanding human diversity in lifestyles and as the foundation concept of their discipline.[6]

Boas and his students stressed studying the "wholeness of a single culture" and "the ramifications of the functional interdependence of its parts for human behavior and diversity."[7] A basic premise of the Boas model for ethnography was understanding the "interrelationship of the parts of one culture instead of cross-cultural comparison to establish evolutionary sequence."[8]

American Indian nations were studied intensely during this period (1880–1930) by many anthropologists but the significant difference, by those who aligned with Boas, was the perspective utilized:

> [I]n light of a cultural relativity and pluralism that denied explicit or implicit judgments on the moral superiority of White civilization . . . they researched Native Americans as tribes and as cultures, not as the Indian (but always as they had once lived in the past and not as they lived when the anthropologist visited the tribe).[9]

Although this change certainly was an improvement in terms of understanding diversity and the wholeness and interdependence within a culture, it continued to maintain the "past" image of the American Indian that was acceptable to American society. Since Boas's work took place during the era of the belief that American Indians were a vanishing race, it is not surprising that the focus was on the past.

Ruth Benedict, a follower of Boas, researched the Zuni, Kwakiutl, and Dobu cultures in 1934.

> She hoped to persuade her reader that the integrative holism of individual cultures was as important as their diversity and that the psychological sources of another culture were as valid as those of one's own. In arguing that one should not make a moral judgment upon the quality of another culture, she, in fact, criticized White American culture through her interpretations of the Dobu and Kwakiutl cultures.[10]

Although Boas and his followers provided another paradigm with which to view cultures, they continued the concept of Indians as "past" or kept the "romantic" vision of Indians, which was, and continues to be, more comfortable for society. Boas developed the theoretical framework for diversity and understanding cultures as a unique and interrelated whole. He promoted the idea of viewing cultures as the people of the culture view themselves. He specifically pointed out the dangers of examining other cultures using the "ethnocentric lens of Western culture," which had been unquestioned to that point in time. But the majority or popular culture of America did not accept or embrace his theories. The Boas tradition did not translate, or make the leap, to American textbooks or education, as was obvious in chapter 1.

In 1928, the Meriam Report identified a major source of the American Indian "problem" as the need for education, which would enable them to help themselves. The recommendations of the Meriam Report were to become part of the New Deal under John Collier who drafted a bill of reforms in 1934. The education section of the bill stated:

> It is hereby declared to be the purpose and policy of Congress to promote the study of Indian civilization and preserve and develop the special cultural contributions and achievements of such civilization . . . [11]

This short statement actually acknowledges American Indians as having "civilization" and "achievements" that should be preserved! However, the bill, which became the Indian Reorganization Act of 1934, was severely cut and the above hoped-for change in educational direction was excluded. Diversity did not become the standard curriculum. The evolutionary scale became the standard curriculum in this country.

How the Theories Became the Standard Curriculum

This section provides an overview of the various definitions of curriculum, the underlying assumptions of social studies curricula, and a paradigm for integrating ethnic content.

Defining Curriculum and Curriculum Theories

Curriculum has been defined as the overall design, plan, philosophy, or set of guiding principles about what is taught and why. Curriculum has been described as including a scope and sequence chart; as a syllabus including rationale, implementation, and evaluation; as content to be studied; as a textbook; as a series of courses; or as a set of planned experiences.[12] Posner defines curriculum as: "An organized set of intended learning outcomes presumed to lead to the achievement of educational goals," and actual curriculum development as the process of selecting and organizing the intended learning outcomes.[13]

Posner identifies several layers of curricula, which a school system implements at the same time. The *official curriculum* is contained in the formal documents endorsed by the state or school district. The *operational curriculum* is "embodied in actual teaching practices and tests." The *hidden curriculum* consists of "norms and values not openly acknowledged by teachers or school officials." This includes socialization, gender relations, issues of authority, and what is valued within the institution. The *null curriculum* refers to what is "*not* taught." The *extra curriculum* includes school planned experiences outside the formal curriculum.[14]

The Underlying Assumptions of Social Studies Curricula

Societal values determine *why* something is important to study, and it then becomes an underlying assumption or theoretical foundation of curriculum. In the United States, one of these underlying assumptions is the goal of producing good citizens who adhere to a homogeneous national identity. One of the primary purposes of the official curriculum is to transmit the nation's goals to the nation's youth.

Schools in the colonial era were designed to create and transmit "a specifically national and uniform culture." It was assumed that immigrants would "cease to be Europeans and become Americans."[15] The same nationalizing standard was applied to American Indians and Africans. Socialization, social control, and assimilation were overt educational goals designed to suppress diversity, thus creating a unified monoculture known as the American national culture. After the Civil War, training was designed to "standardize the product" to provide workers for the industrial society. Vallance maintains that the overt goals of homogeneity, social order, standardization, and conformity became the hidden curriculum around the turn of the twentieth century, because these American values had become ingrained within educational institutions.[16]

From the outset, social studies as a topic of study combined history, geography, economics, and sociology. In the early 1900s, Hampton Institute, under the direction of Thomas Jesse Jones,

> evolved a new social studies . . . that was designed to equip America's underclass with the skills that would bring them to the level of the white middle class.[17]

This experimental curriculum was tested on freed slaves and American Indian youth at Hampton Institute who, at the time, were viewed in Morgan's scale as in earlier stages of development.

Jones chaired the National Education Association subcommittee on social studies which drafted the guidelines changing social studies from the study of history to the production of good citizens. In 1916, Jones's perspective on social studies as a design to produce citizenship, based on the Hampton experience, became part of the operational curriculum when it was endorsed by the National Educational Association.[18]

In creating a uniform national identity, studying diverse cultures was *not* a goal and so became part of the null curriculum, or what is not taught. Thus, the study of cultures was ignored or distorted in mainstream curricula that focused on the history of great American heroes and great wars.[19] Cultures were perceived to be on a hierarchical scale, with Western civilization, or the American national identity, at the top and all other cultures placed at a lower level or at the bottom of the scale. Scientific researchers, including Boas, measured the size of skulls to ascertain brain capacity in the effort to prove that northern Europeans and their descendants were superior to American Indians and African Americans. These cultures were labeled as "less than," "exotic," or "vanishing" and were included in curricula only on a superficial level or became part of the null curriculum.

The study of various cultures has, in most cases, been limited to geographical location and artifacts. Banks describes this approach as:

> the common, isolated use of artifacts and other aspects of the material culture of people, without a holistic interpretation, as the "museum approach." It reinforces the "us"-"them" difference and highlights a "hierarchy of cultures" based on the way the outsider perceives the minority.[20]

Such an approach is replete with stereotypes, inaccuracies, and distortions regarding the cultural heritage of this nation. The "hierarchy of cultures" became part of the hidden curriculum.

The standard scope and sequence for social studies begins with the study of the individual, family, community, state, nation, and world as each relates to the national identity.[21] In this standard model, cultures often become either part of the null curriculum or are included only as artifacts or as exotic entities.

With the civil rights movement and demands for equity and inclusion, textbooks began to include special boxed information on famous people such as Martin Luther King Jr. or Chief Joseph.

Banks has identified this approach as the contributions approach (see Figure 3.1). By including this information in separate boxes it remains outside the main body of the text, therefore outside of the mainstream history. Another change has been the celebration of holidays and the development of units on specific cultures, or the additive approach. Again, these units are outside mainstream history. These measures have been an improvement, but the standard mainstream curriculum with its underlying assumption of producing homogeneous American citizens continues, thus denying the cultural diversity of this nation. The national goals are ever so slowly beginning to move toward recognition of the cultural diversity of this nation. The debate over this change will continue well into the future.

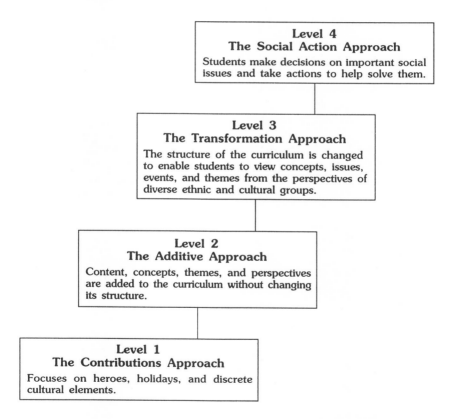

Figure 3.1. Levels of Integration of Ethnic Content (from: James A. Banks, *Multicultural Education,* 1989:192)

A Model for Integration of Ethnic Content

The theoretical perspective for the culture-based model presented in this text emerges from Banks's framework for identifying the levels of integration of ethnic content into the curriculum (see Figure 3.1) In this country, we have only witnessed levels one and two of his model. In level three, the transformation approach,

> (t)he structure of the curriculum is changed to enable students to view concepts, issues, events, and themes *from the perspectives of diverse ethnic and cultural groups.*[22] (emphasis added)

The transformation approach promotes *understandings* about diverse cultures by *including multiple perspectives*. Restructuring curriculum to present diverse perspectives beyond the level of the trivial or the level of addition to the mainstream curriculum should be at the heart of curriculum. Multicultural education as a movement has tended to focus on the political issue of equity, that is, equal rights, for ethnic and minority students.

This study is based on understanding culture and understanding diversity, and proposes a conceptual framework for implementing Banks's transformation approach. The culture-based approach to curriculum provides a way to restructure the traditional curriculum by recognizing, valuing, and including diverse cultures. In this nation people have layers of identity based on their cultural heritage.

Chapter Summary

Although Boas provided another perspective, and Collier tried to recognize American Indian civilization, neither reform was to make an impact on how American Indians are presented in the educational system of this country. The cultural evolution scales, the stereotypes, the omissions and misinformation continue to plague American educational materials. The multicultural approach to integrating ethnic content has been able to implement only level one, contributions, and level two, additive approaches. Thus, it is imperative that we begin to create new models that acknowledge the perspectives of specific cultural viewpoints, thereby moving into the transformation level.

The culture-based framework, described in the next chapter, provides a starting point to broaden our knowledge base and move toward understanding and accepting diversity. This new framework, applied to educational materials, provides a way to present multicultural perspectives through curriculum development. Chapters 5 through 9 contain the case study on the Haudenosaunee with the thematic focus on corn, which provides the background information necessary to present the perspective that the "wholeness" of a culture from the specific cultural perspective must be integrated into curriculum.

Chapter 4

Valuing Diversity through Culture-Based Curriculum

To move away from the theories that support hierarchical ranking of cultures requires readjusting our thinking by moving into a circular framework which examines the many facets of a specific culture. To understand the interrelationship, the interdependence, the "wholeness" of a culture demands that we think in other than hierarchical/evolutionary terms. The underlying assumptions in this framework are:

1. All cultures have value.

2. All cultures have a world view that structures their values and their society.

3. All cultures have an indigenous knowledge base.

4. All cultures are dynamic.

How do we develop a framework within which cultures are valued? In order to develop respect for indigenous knowledge contained within diverse cultures there are three underlying principles that must be recognized. Barreiro stated these principles in regard to publishing an indigenous peoples journal, *Northeast Indian Quarterly,* and these basic principles are definitely applicable to curriculum.[1]

1. The first principle acknowledges the existence of Native People's intelligence, meaning that the traditional ancient knowledge base of indigenous peoples continues to exist and have value within the context of the modern world.

2. The second principle encourages and respects self-definition by Native Peoples.

3. The third principle encourages the celebration of distinctiveness, of diversity, in this modern world with its multicultural reality.

With those thoughts in mind, we can proceed to look at an alternative way of viewing many cultures. One perspective which everyone can accept is that we are all human beings, and as human beings we seek ways to understand and order our lives.[2] The All Cultures Chart (Figure 4.1) illustrates the basic components all human beings utilize to understand and structure their world.

The cultural components, which apply to every culture of the world, are shown in the All Cultures chart (Figure 4.1). This chart

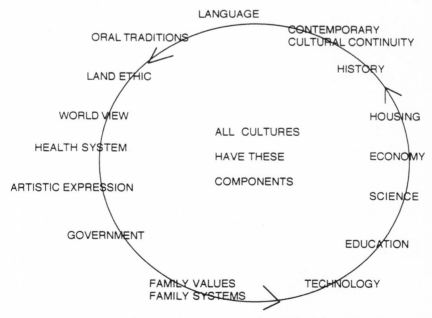

Figure 4.1. All Cultures Chart: Cultural Components of All Cultures

can be applied to the many diverse cultures around the world. Because this chart is circular there does not exist a higher or lower culture, which means that value judgments are not to be made. We may not be able to understand why another culture does specific things differently, but we are to understand that the members of that culture do understand and *have the right to practice their beliefs without judgment by outsiders* who can't completely understand their culture because they haven't grown up and spent all of their lives living within that culture.

All human beings have language, and until the printing press, all cultures relied on oral traditions to transmit their cultural values to the next generation. Land ethics are prescribed within the world view of the culture. Does the culture value land and understand land as a gift to be carefully utilized and preserved by human beings, or does the culture fear nature and seek to control the natural environment? The stories within the oral tradition explain and transmit the values and knowledge that compose the world view of a specific culture. All cultures have a health system that includes knowledge of plants as food and medicines for healing, and practice wholistic healing, which includes the physical, spiritual, mental, and emotional well-being of the person. Every culture has unique artistic forms of expression including their culturally specific forms of music, dance, drama, and literature. Whenever two people or more live together they develop some form of government with rules to guide human behavior. The type of government varies widely depending on the culture. The oral traditions contain the essence of valued and acceptable behavior within that culture's government. Family structures may vary widely across cultures, including clans, extended families, and nuclear families, but all cultures have family values. Technology was developed throughout time because all human beings strive to find an easier way to accomplish work. Education has taken many forms throughout time and across cultures because all cultures transmit their knowledge to the next generation. Human beings are naturally inquisitive and have made observations of natural phenomena and engaged in scientific experiments to answer their questions. Every culture has developed some form of economy to support its population. Housing styles differ greatly but all cultures have created housing to suit their needs.

What about changes over time and changes when people from different cultures meet? All of the components in the above chart comprise the foundation of the culture. No culture in this world is static. All cultures are dynamic, but their basic values, their world view maintains its integrity and beliefs. Components within the All Cultures Chart change within dynamic patterns to meet the people's needs across time and events. The changes include cultural adaptation to new technologies, such as speaking the native language over the telephone, or putting beads on baseball hats, but the basic beliefs continue to guide the culture.

You may ask, what about history? Every culture's history began with their creation story and their history has continued ever since. Interactions with other cultures, before and since the written word, have impacted cultures. These interactions have had both positive and negative impacts on the culture, but the fact that so many unique cultures continue to exist attests to their cultural continuity across historical events. Contemporary culture includes every aspect of the All Cultures Chart; the enactment of the culture may have adapted, that is, changed, over time, but the basic world view of the specific culture continues to guide contemporary enactment of the cultural beliefs and values.

In Figure 4.2, these universal human concepts have been applied to the Haudenosaunee who had, and continue to have, the universal human elements necessary for perpetuation of their world view. The differences in this specific model are unique to the Haudenosaunee. When we apply this model to any of the more than 300 North American Indian nations that continue to exist

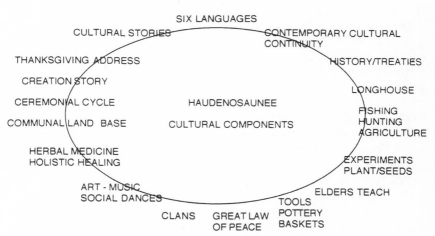

Figure 4.2. Haudenosaunee Cultural Components

within the United States, we can identify specific differences that are based on the uniqueness of each nation. For example, languages are different, ceremonial practices differ, the economic base will differ based on available resources, but there is language, world view, ceremony, and an economic base. This chart will be examined in depth in chapters 5 through 9.

Culture-Based Curriculum Framework

The culture-based curriculum framework developed in this study provides a place to begin the process of developing curriculum that includes and values diversity.

THEORETICAL FOUNDATION	PRACTICAL APPLICATION
A	E
Underlying Assumptions Valuing Cultural Diversity Holistic: Culturally Specific Paradigm Living History/Dynamic Cultures	*Criteria for Developing* *Culture-Based Curriculum* Multiple Perspectives The Team Approach
B	F
Cultural Research World View Way of Life Interaction of Cultures Dynamic (Selective Adoption) Cultural Continuity/Living History	*Interdisciplinary Approach* Teaching Across Cultures
C	G
The Interrelated Components of the *Culturally Specific Paradigm*	*Instructional Approach* Culturally-Based Materials Teaching Strategies: Combining
D	Cultural Research (see B) and Thematic Focus (see D)
Thematic Focus Identifying a major theme from the culture	Evaluation

---

Figure 4.3. Culture-Based Curriculum Framework

The outline for the culture-based curriculum framework is shown in Figure 4.3. Items A through D form the theoretical basis for culture-based curriculum. Items E through G provide insight into the curriculum development process utilizing the culture-based framework. This framework provides a way to think about and process the cultural information necessary to design culture-based curriculum.

A: Underlying Assumptions

If we are to recognize the fact that America is a land of diversity then the definition of citizenship will change from the homogeneous model to a view of producing citizens who value the cultural diversity of this nation. This country would move toward Ellison's view of Americans as having layers of cultural identity. The curriculum models would change from exclusion to inclusion of many cultures, and would strive to present the value and integrity of diverse cultures. How we view diversity ought to change from fragmented stereotypical views to studying cultures as holistic entities.

Three major concepts form the underlying assumptions, the foundation, of the culture-based curriculum framework. First, the concept of cultural diversity establishes the perspective with which to view diverse cultures. Second, presenting diverse cultures as holistic culturally specific paradigms provides an in-depth view of a culture. Third, delving into the living history of the culture presents the contemporary realities that are based on cultural beliefs and historical events.

Valuing Cultural Diversity

The basic assumption of the culture-based curriculum framework recognizes that all cultures have fundamental value and integrity.

> Boas's fundamental criticisms of the evolutionary anthropology of his time concerned method and temperament. In method, Boas advocated a holistic, detailed, and exhaustive ethnography of specific tribes. In temperament, he was a proponent of cultural relativism, a nonjudgmental and empathetic attitude that recognized the fundamental value and integrity of all cultures, primitive or civilized. Thus Boas sought an anthropology with both scientific rigor and humanism, which were lacking in much of the evolutionary theory of the period.[3]

Note that Hanson adds the phrase "primitive or civilized" after the definition of cultural relativism, then criticizes the evolutionary theory he has just reinforced! This example shows the complexity of moving from the standard evolutionary theories to accepting diversity. It is a struggle that continues to the present day. Omitting that phrase, and using the concept of "value and integrity of all cultures" lays the foundation for the culture-based curriculum framework.

The western hemisphere was a land of diversity long before the Columbus encounter. It has remained a land composed of unique cultures despite the attempt at homogenization.

Holistic perspective

The next step, after accepting that diverse cultures have value and integrity includes accepting differences by realizing that those differences are pieces of a holistic culture. Each culture carries out its world view and way of life, differently—not less than, nor on a lower level—just differently. We are not to judge those differences, nor do we have to agree with those differences, but we must come to an understanding that every culture is unique and has the right to exist. Within each culture those behaviors that we may see as different make perfect sense when understood with the holistic structure of their way of life, which forms their culturally specific paradigm. Conceptualizing diverse cultures as holistic enables us to move beyond the "artifact" approach. Providing a holistic view of diverse cultures allows instruction to move away from stereotypes, artifacts, contributions, and additive approaches.

Living history/dynamic cultures

The third underlying assumption maintains that all cultures are dynamic, that is, fluid and changing. We can find evidence of that dynamic nature in the contemporary, or living history, in which the world view of the culture continues to exist and guide contemporary lives and history. It is important to remember that cultures were dynamic before contact as well. Selective adoption, the two-way exchange of cultural elements, continues to occur within the contemporary time period. This exchange happens in two directions whether we're discussing philosophies of democratic government, environmental issues, or using metal kettles instead of clay pots. Living history provides the opportunity to dispel the "primitive," "savage," and "vanishing race" stereotypes.

Living history requires interaction with Elders by asking them to share their knowledge. Through this interaction young people will have the opportunity to gain valuable knowledge and respect for our old ones. The underlying assumption in this model maintains that the Elders possess the culture's world view. It is the contemporary Elders' cultural knowledge, combined with written archival sources from the late 1800s through the early 1900s, that provides a broader perspective. Living history is evident in current issues and events that can be examined to determine how the clash of world views directly impacts present-day problems and conflicts.

B: Cultural Research

Culture-based curriculum supports a broader understanding of other cultures by providing information to students within a framework that values diversity.

Therefore, culture-based curriculum requires study of:

1. the world view expressed by a particular culture,

2. the way the world view structures the people's way of life, and enactment in the daily life of the culture,

3. the interaction of cultures,

4. the dynamic aspects of culture, selective adoption/adaptations, and continuity of world view, and

5. cultural continuity to contemporary times.

World view

World view contains the knowledge, the philosophical base, with which a culture defines itself. World view contains the beliefs that form the central core of the culture's understanding of the world, and world view shapes the cultural enactment of those beliefs. That world view is structured within a "culturally-specific paradigm for viewing the universe."[4] The "culturally-specific paradigm" guides culture-based curriculum development.

Geertz defines cultural world view as the "picture of the way things in sheer actuality are, their concept of nature, of self, of society." He explains that world view contains the culture's "most comprehensive ideas of order" and defines a "way of life" that is "an authentic expression" of the world view.[5]

Thus, the world view of a culture is composed of many elements woven together and only through looking at the interconnectedness of those elements can we begin to grasp the meanings implicit in the world view. It is through examining in-depth descriptions to understand the symbolic meanings of a culture that we begin to develop a holistic perspective of a culture different from our own. Anthropological interpretations have reduced the complex cultural elements into rules, tables, and charts, in essence formulas, that lose the interconnectedness of the culture. When a teacher has students do trivial activities such as string macaroni beads for wampum belts, or build cardboard longhouses, students are left unaware of the intricate ways these items function within the holistic Haudenosaunee (Iroquois) way of life.

Alfonso Ortiz, a Tewa anthropologist from San Juan Pueblo, provides a definition of world view in the context of land and tradition:

> World view provides people with a distinctive set of values, an identity, a feeling of rootedness, of belonging to a time and a place, and a felt sense of continuity with a tradition which transcends the experience of a single lifetime, a tradition which may be said to transcend even time.[6]

Understanding the importance of that "rootedness" will further the understanding of why a people defend their homelands. Ortiz further defines world view as "a cultural system in the sense that it denotes a system of symbols by means of which a people impose *meaning* and *order* in their world."[7] (emphasis added) The "feeling of rootedness" refers to the land base that each Indigenous People identifies as their place of origin and continuity. Whether the people are Tewa Pueblo, Onondaga, Lakota, Maya, or another Indigenous People, they have a connection, a "rootedness," with a particular area and a specific cosmology.

Myths, legends, and epic narratives contain the essence of a people's world view. That world view establishes a way of life for the community. The community enacts world view by conducting the yearly cycle of renewal ceremonies and maintaining the beliefs and values contained in stories of origins and "dramatic narratives."[8]

The world view of a culture is contained in its cosmology, myths, legends, folk heroes, and epic dramas, which, until quite recent times, have been handed down to each generation through the oral traditions. These primary sources explain and give meaning to the culture's origins, the interrelationship and interdependence of nature and humans, the way of life, the form of government, and the way it adapts the old ways to the newer realities.

In striving to understand other cultures, one must search for those elements or themes in the culture that express meanings that are enacted in daily life. What do people within the culture see as truth, as reality, as instructive, as a way of life? What are the "cultural systems of knowledge" that guide the culture? What are the themes "which illustrate the most general unifying principles of . . . existence"?[9]

The world view of a people provides the basis for human belief and behavior. The world view establishes the community consciousness, and in this way enables human beings to live together in structured societies. Vecsey says myths that contain world view are the

> means to confer life, promote life. The lessons they teach tell not only how the world came into being, but also how humans can survive in the present order of life. They are pragmatic as well as expressive narrative that people use in their everyday lives.[10]

To look at what Ortiz calls "cultural systems of knowledge" and what Sosa calls "culturally-specific paradigm" it is necessary to identify recurring themes and values in the world view. One must understand that these themes are intertwined and interdependent.

Way of Life

How people perceive their world determines how they structure their ceremonies, economy, government, education, technology, and health systems. (see Figure 4.4). In order to make the abstract concepts of "world view," and "way of life," into concrete terms usable for instruction, it is imperative to select a thematic focus from the culture. The case study in this text focuses on corn as the thematic focus because corn can be used to illustrate complex interrelationships within Haudenosaunee culture, as well as interactions between cultures, and allows these relationships to be integrated into instructional units.

Each of these cultural components relies on the other components and together they form the "cultural systems of knowledge" and the enactment of the "culturally-specific paradigm." Therefore, what has previously been viewed as artifacts, or parts of a culture, are interwoven into a complex whole within the culture. To under-

stand the complexity of a culture, one must view the cultural information in a way that demonstrates the interrelationships of each
aspect of a particular culture.

Interaction of Cultures

Cultures interact on many levels, including the two-way exchange
of knowledge, technology, and food, as well as in conflicts over land
resources and power. The purpose of looking at these cultural interactions is to understand both the positive and the negative effects on both sides of the interaction.

For example, there are fundamental similarities in world view
and the cultural enactment of world view in Mayan, Hopi, and
Tewa Pueblo cultures, among others. Because corn is indigenous to
the western hemisphere, there are similarities across cultures in
this hemisphere, although there are distinct differences in language, ceremonies, dress, housing, and land base.

In the early days of contact there was an enormous exchange
of technology and trading networks between the civilizations of the
Old World and the New World.[11] Indigenous Peoples of the northeast woodlands shared their vast agricultural knowledge and technology with the new arrivals from Europe who depended on
American Indians for their very survival in the early years.

Historical documents provide information on the important
cross-cultural exchanges of knowledge and events. Often these
dramatic interactions were affected by the differing world views of
each culture. It is important to study the clash of world views
behind these interactions in order to understand these exchanges
and conflicts. Textbooks often present only one side of the interac-
tion/conflict. The culture-based curriculum model proposes looking
at interactions from many perspectives to broaden our understanding of the events.

Dynamic Aspects of Culture—Selective Adoption

Over time all cultures change. All cultures affect, and are affected
by, the forces of the times. In his research on Maya cosmology, Sosa
offers a way to understand the old and new.

> Whether we refer to Maya Cosmology then, as their "world
> view" or a "theory of the universe," we are acknowledging that
> it is a systemic paradigm which an individual can draw on to
> understand new, incoming information.[12]

These adaptations, keeping elements of the old while dealing with current reality, make a culture dynamic. Replacing a stone hoe with metal hoe does not change the basic world view. However, if the culture ceases to pass on the oral traditions to the next generation there would be a dramatic change. The influx of printing press, telephones, television, and the mass media has affected all oral cultures. However, in spite of the impact of mass media the world view continues. The stories, the myths, the dramatic narratives are now written down and published in books. Thus, young people can read the information as well as hear it orally, but they must go to the Elders, the culture bearers, to hear more, to ask questions, to verify knowledge, and seek a deeper level of understanding. The culture continues, with adaptations, but the basic world view has not changed. People from all cultures conduct their lives in the present world reality, as contemporary persons, with their traditional world view continuing to define the parameters for what is appropriate. The world view gives the young people the strong identity required to cope with the modern world.

Selective adoption is the process of taking an element from another culture and adapting it to fit. Ray Gonyea, an Onondaga, describes this process:

> This period [Civil War to World War I] began with the Onondaga Nation's adopting selective elements of the non-Indian world into their culture, such as clothing and farm equipment, but always retaining a core of their culture that was distinctly Iroquois.[13]

Understanding the intricate process of selectively adopting something and then adapting or incorporating it into the cultural world view demonstrates adaptation and continuity of the culturally specific paradigm. Adaptation does not mean the old way is diluted, changed, forgotten, or is less than what it was. Adaptation means there may be a new, or different way, of expressing the original in the context of the contemporary.

Cultural Continuity/Living History

Because cultures are dynamic they change and selectively adapt items from other cultures, but they maintain their world view. Studying a culture includes a contemporary view of that culture so that students are not left with misperceptions associated with the "vanishing race" stereotype.

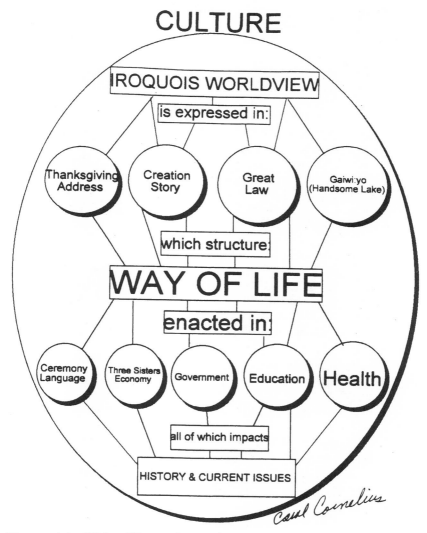

Figure 4.4. Living History Approach

The world view of the specific culture affects how that culture has interacted, and currently interacts, with other cultures. The living history approach to studying cultures (Figure 4.4) provides a way to understand how, and why a culture responds to current issues in a specific manner. Whether the issue is environmental, political, economic, or social, the culture provides a way to approach the issue. Cross-cultural understanding can be enhanced by under-standing the culturally specific perspective guiding people's view of a current issue. What we see in the daily newspapers as conflicts are directly the result of the clashing of world views. The living history, the contemporary life, of a people is based on the continu-ity of the world view, interactions with other cultures, and main-taining the culturally specific paradigm within the modern world. To understand the perspectives of another culture requires a com-prehensive knowledge base from which to examine how the world view forms the basis for contemporary action on issues and events.

As in any culture, those specifically designated people who study the beliefs and keep those beliefs alive are also the people who teach the next generation. The Elders, the culture bearers, transmit the world view to the next generation.

> But more importantly, the old were the community's most re-spected source of information and opinion. For the old were the people who had lived the longest and who had seen or experi-enced the most in life. They were also the teachers, who handed down the traditions of the past to the younger generation. They were in a sense the villages' library or encyclopedia.[14]

Hammell was describing a "prehistoric" Seneca community in the above statement, and the same respect for the Elders continues today.

Many cultures are fortunate to still have those Elders who retain the cultural systems of knowledge, the culturally specific paradigms. They share and teach the cultural knowledge that pro-vides identity and the "feeling of rootedness . . . the . . . sense of continuity with a tradition that transcends the experience of a single lifetime, a tradition which may be said to transcend even time."[15]

C: The Interrelated Components of the Culturally Specific Paradigm

Looking at the culturally specific paradigm illustrates the interre-lated beliefs that are significant to a culture and provides a holistic view of that culture (see Figures 4.1, 4.2, 4.4) The following ques-tions are a guideline for looking at a culture to identify the essen-tial components: How does the culture view itself? What's important to this specific culture? What does this culture value? How do the beliefs structure the way of life? What are the significant cultural themes? How did this culture interact with similar cultures? How did this culture interact with different cultures in the early historic contact period? What evidence is there of selective adoption? What evidence shows that the interaction of cultures was a two-way exchange? How has this culture maintained the continuity of its world view and way of life into contemporary times? What do the people of the culture want taught about themselves?

When you can see how the oral traditions of a culture define their land ethic which guides their economy, which connects with the ceremonial cycle, child-rearing practices, and music, dance, songs, all of which circle back to the oral traditions, then you can understand how very culturally specific each culture has been, and continues to be.

Cultural exchange has always taken place easily between similar cultures. However, when contact took place between two dramatically different cultures, within a significant historical event, there were both positive and negative exchanges of knowledge. For example, a two-way exchange of knowledge began with contact between the western and eastern hemispheres. Whether interaction takes place with similar or dramatically different world views, each culture maintains its own world view. The dynamic nature of all cultures means that every culture continues to change in some ways, yet the foundation, the world view, the very essence of the culture, maintains the cultural values within the culturally specific paradigm.

The underlying assumptions and cultural research framework provide a holistic way to view another culture instead of using hierarchical scales, stereotypes, and value judgments. This model allows an examination of a complex culture with respect and dignity. It examines cultures on a continuum of past, present, and future because they have not disappeared, and these cultures continue to be viable in the contemporary world. Application of this culture-based framework and a thematic focus to curriculum development will provide a broader base of information to develop instructional materials on diverse cultures.

D: Thematic Focus

Each culture has a specific set of beliefs that explain how this world came to be and how human beings should live on this earth. This set of beliefs forms the foundation of their culturally specific paradigm. Within that paradigm are a series of interrelated factors, a world view, that structure this culture's way of life and how they interact with other cultures. The world view contains an interconnected web of relationships necessary to maintain the culture's world view (see Figures 4.1, 4.2, 4.4).

By examining the world view and related cultural components, a significant element of the culture emerges. A thematic focus emerges as you experience the process of trying to understand a culture different from your own. Themes such as corn, as used in the case study, or music, a drum, a rattle, are concrete items that are utilized in the culture-based curriculum framework to show the connection to all the elements of the All Cultures Chart. Instead of isolating an item from its culture, *the item becomes the center of the circle and connects to all the components.* The thematic focus includes not only the specific topic, but the dynamic nature of that topic within the culture and the interaction between cultures. The selected element becomes the thematic focus around which the curriculum is organized. Instructional plans revolve around the thematic focus. For example, in the Haudenosaunee case study corn became the thematic focus because of its centrality to many aspects of the Haudenosaunee way of life, and corn can be studied today as an important component of the Haudenosaunee culturally specific world view. The thematic focus emerges from the culture because it is an important central aspect of the culture, of the way people understand the world to be.

How does one go about identifying a thematic focus? Begin by asking people from the culture what they think should be taught about their culture in the public schools. The thematic focus is a significant component of the interconnected web of relationships within the culture, therefore it can be shown how the item or concept connects to the holistic world view. An aspect of the culture becomes an artifact only when it is disconnected from its place within the web of relationships and taught separately. Using a thematic focus provides a way to counteract stereotypes of the culture by examining specific cultures as holistic entities.

The thematic focus can also be studied cross-culturally by comparing similarities and differences across cultures. An example with corn would compare ceremonial practices, stories, planting techniques, technology, foods, and storage methods. This approach allows a teacher to take an artifact or isolated item of a culture and expand it to provide a holistic view of the culture. For example, in the artifact approach using a chart that shows styles of housing from American Indian cultures ends with students thinking American Indians still live in teepees, thus avoiding contemporary realities. In the culture-based approach, the teacher and students would examine how and why a particular culture utilized a specific type of housing based on environmental conditions and cultural beliefs and practices, and the unit would include contemporary housing styles. Oral traditions which establish values, are highlighted, artistic symbols are identified, and the social structure of the community is examined. Questions could be asked such as how specific housing structures reflect the family values, because in many cultures the community ethic revolves around the concept of the entire community being involved with assisting each other within extended family or clan structures. To expand and explain interconnecting relationships we should ask, how do the oral traditions relate to family structures (kinship, clans, extended family) and economy (three sisters agriculture) and science (hill planting, fertilizers, cross-pollination, nitrogen cycle)? Another example would be to examine how in the northwest salmon forms an integral link between economy, world view, ceremonies, way of life, and social organization. The same pattern can be applied to wild rice in the Ojibwa culture of the northern Great Lakes.

As the curriculum develops, the instructional plan constantly connects the thematic focus with its relationship to the holistic cultural components (see Figure 4.1). The instructional objectives reflect the connection between the thematic focus and the culturally specific paradigm. Given the time constraints and workload of teachers it is important for the curriculum team to identify a thematic focus and develop a circular chart of the culturally specific paradigm that serves as a unifying center for the curriculum. Consistently verifying the information gathered on the components of the All Cultures chart with the culture bearers assures the use of accurate information.

E: Criteria for Developing Culture-Based Curricula

Once we have a clear understanding of the conceptual framework for culture-based curricula we can begin the process of developing curriculum (see Figure 4.3). The four criteria for developing culture-based curriculum are:

1. presenting multiple perspectives through a team approach,

2. selecting a thematic focus that emerges from the culturally specific paradigm. (see D)

3. utilizing an interdisciplinary approach (see F), and

4. recognizing continuity of the culture into the contemporary era (see G, Teaching Strategies).

Multiple Perspectives—The Team Approach

Once the school team has agreed upon the culture, or cultures, they plan to study it is imperative they contact people from the culture who are the the best source of cultural knowledge. These culture bearers then become part of the team. This team must include culture bearers, that is, the Elders, who can share the culturally specific paradigm of their people.

Between Sacred Mountains, Navajo Stories and Lessons from the Land describes the team of people who created a curriculum for Navajo students. The team of developers are described as: Storytellers and Teachers, Seekers Who Asked and Understood, Artists Who Looked and Drew, and Listeners, Learners, and Scribes. The document also acknowledges the Rock Point School Board and Parent Committee.[16] The description of the team indicates this group of curriculum developers had a different perspective on "experts" and expertise than the standard curriculum that generally looks to, and acknowledges, only curriculum developers, teachers, and administrators as "experts." Another example of the team approach lists, "Elders, tribal leaders, historians, educators, parents, and students."[17] This listing includes representatives from the perspectives of teacher, learner, and subject matter.[18] Including the Elders and tribal leaders expands the definition of "expert" by including the culture bearers as significant, and legitimate, sources of knowledge. This would seem to be just common sense to people within the culture, but remember academia has not recognized the expertise of the culture bearers other than as mere informants. Therefore the team must include the cultural community not only to gather cultural information, but also as the authority on that information. Perhaps the best way to describe this team approach to curriculum development is to say that it develops expertise from the bottom up rather than from the top down. The team approach values many forms of knowledge.

Contacts with people from the culture can usually be made through educational programs or offices directly operated by the culture. When curriculum developers use only printed resources they encounter the danger of having only the limited *interpretation* of the so-called experts who, most often, are not from the culture.

The team then begins the process of gathering information about the culture. Historical documents, oral traditions, newspapers, and magazine publications by the culture, archival documents and pictures, old newspaper articles, and most importantly interviews of contemporary elders and educators provide rich sources of materials.

F: Interdisciplinary Approach: Teaching Across Cultures

Using an interdisciplinary approach to teach about cultures is essential because the components of culture are interdependent. The team can avoid the "artifact" approach by using instructional methods that present the thematic focus as connected to all aspects of a culture. If components of a culture are taught in isolated subject matter categories such as social studies, literature, or art, students receive a fragmented view of that culture instead of an integrated holistic view. Thus, students are not presented the culture in a way that leads to understanding the integral relationships within the entire culture. The whole language approach to reading provides an example of an interdisciplinary method.

One of the best examples of utilizing the culture-based and thematic focus to go beyond the artifact approach was explained by an Onondaga Elder. He outlined using corn as an instructional unit designed across: culture, art, math, creative writing, and language, both Onondaga and English. Culture, he explained, would include the ceremonial cycle, stories, songs, dances, and foods made with the corn. Many aspects of the Onondaga language could be incorporated into the unit for Onondaga students. He envisioned the students using artistic skills to illustrate stories, writing stories in English to improve writing skills, and creating plays. Math could be incorporated in studying the productivity of one ear of corn. This study would include science to study the genetics of corn, and history would be studied to trace the continuity of using white flint corn within the Onondaga Nation Territory. This Onondaga elder applied the same culturally holistic style of teaching around a thematic focus to study making baskets, or a water drum, thus showing how these items are essential components of the holistic culture (see appendix).

Students can begin to make cross-cultural comparisons of similarities and differences. One of the benefits of this approach is that it provides a way for students to begin valuing many cultures by identifying and understanding similarities. For example, to provide

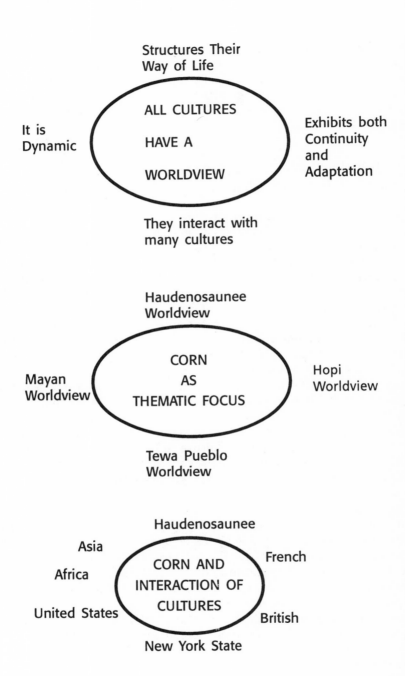

Figure 4.5. Cross-Cultural Studies

students with diverse perspectives they must understand that: all cultures have a world view; the world view structures their way of life; the culture is dynamic; the culture interacts with many cultures; and the culture exhibits both continuity and adaptation to the present day. Figure 4.5 illustrates a flow from understanding how all cultures have a world view, to using corn as the specific thematic focus to examine Mayan, Tewa Pueblo, and Hopi world view. Next, corn as the thematic focus provides a way to look at the interaction of many cultures and time periods. Developing cross-cultural studies allows sophisticated comparisons across cultures. The implications of culture-based curricula, with the emphasis on inclusion of many cultural perspectives, would allow students to view cultural diversity, not as a liability, but as an asset, and provide a way to expand instruction in a holistic way.

The interdisciplinary nature of culture-based curriculum involves utilizing the thematic focus across science, art, math, creative writing, reading, drama, music, economics, history, sociology, and language to develop standard social studies skills.

G: Instructional Plan

Utilizing Culture-Based Materials

In some school districts and tribal education programs the indigenous people have recorded the oral traditions of a specific culture for local usage. However, as often happens in the public school system, these materials have been used as additive units rather than being incorporated throughout the standard curriculum. The culture-based curriculum framework promotes integration of culturally based materials, including an interdisciplinary approach, across the curriculum. These culture-based, and culture-produced materials provide a cultural perspective from within the culture.

Locating these materials can be a difficult task. The Elders and people from the culture who work within the educational context are excellent sources of information on these materials. Asking people from the culture to speak in your classroom continues to be the best resource. When students hear a story told by a cultural person, they become totally engaged in the story. One word of caution, however: Culture-based curriculum means that the study of another culture is not just one day or one week or one speaker, it is a continuous process of inclusion throughout the school year.

The culture-based framework includes focusing on instructional materials obtained through interviewing the Elders, studying historical documents, analyzing government policies, examining materials produced by the culture, comparing art across several time spans, and locating old and contemporary photographs. Each source of instructional materials utilized provides more information and broader perspectives. Instructional materials such as videotapes, filmstrips, works of art, oral stories, music, dance, and guest speakers provide an exciting way to present a culture to students. The education department for a specific culture provides the best source of information. Regional or state education departments and university education departments may be sources of assistance in locating these items or making the necessary contacts with people who have access to the materials.

Teaching Strategies: Combining Cultural Research and Thematic Focus

Teaching strategies for culture-based curriculum are enhanced by using experiential approaches and inductive inquiry approaches in studying diversity. The experiential approach engages students in activities and projects. Taba's inductive inquiry approach to thinking skills allows the students to interact with the information and form their own comparisons of data.[19] This approach encourages students to develop their own concepts about cultural diversity based on their research. Students can then apply their concepts to find similarities and differences in other cultures, and to understand the interaction of cultures.

Because Haudenosaunee culture consists of an interrelated culturally specific paradigm, an interdisciplinary curriculum design provides the best way to implement instruction. This section provides practical ways to apply the culture-based interdisciplinary approach to the thematic focus on corn.

Many media for instruction are included, such as stories, paintings, photographs, line drawings, maps, time lines, charts, diagrams, biographies, and interviews. Skills are developed across disciplinary lines, including social studies, art, math, and language arts, by developing abilities in creative writing, reading, listening, critical analysis, critiquing art, problem solving, language, and research.

The following examples show ways to utilize the thematic focus of corn using the categories of the culture-based curriculum framework incorporating an interdisciplinary design.

World View—The world view is found in the oral traditions of a culture. As mentioned above, some of these oral traditions have not been recorded and published until quite recently. Contacting the people from the culture is essential in obtaining oral traditions or in verifying published accounts. The first step in trying to appreciate another culture involves finding the epic narratives that are central to the culture and additional oral traditions or stories that the culture bearers identify as important. Have students *discuss the values* expressed in the story and how these values show the *interconnected relationship of humans and nature.*

The Thanksgiving Address shows the ways in which humans, animals, and nature are interconnected in Haudenosaunee world view. Kinship terms the Haudenosaunee use to describe various elements of nature are an important theme to have students recognize as they correlate the Ernest Smith paintings to each aspect of the Thanksgiving Address. Having students illustrate the Thanksgiving Address in their own style can provide insight to their understanding of the major concepts. Do they use a circular pattern to show interconnections within nature or line drawings, or square boxes?

The role of corn within Haudenosaunee culture becomes evident by combining the oral traditions with the Ernest Smith paintings. An analysis of the paintings would include such questions as: What aspect of Haudenosaunee culture does Ernest Smith share with us in each painting? Does the material culture depicted indicate a pre-contact theme, post-contact, or is it a combination? What relationships are depicted? Does this painting add to our understanding of the interconnectedness of human beings, animals, and nature? Have students analyze the continuity of Haudenosaunee world view by looking for significant cultural symbology utilized in contemporary Haudenosaunee art.

Develop students research skills by having them find corn stories, legends, and epic narratives from diverse American Indian cultures throughout this hemisphere. Identifying significant cultural elements in these stories encourages students to compare and contrast similarities and differences. Map skills and geography are incorporated by identifying where these cultures are located.

Way of Life—The ceremonial cycle, and nature's cycle, connect people with their community and the seasonal cycle to maintain culture and community balance. How does the ceremonial cycle show us the importance of corn in Haudenosaunee life? How does the ceremonial cycle include thankfulness for the seasonal cycle of nature and the agricultural cycle?

Corn is dependent on human beings and human beings are dependent on corn, in a reciprocal relationship. Design an experiment by planting a whole ear of corn in one container and in the second container place individual kernels of corn. The whole ear will sprout but the plants will suffocate each other because they need to be separated as in the second container. This experiment will show the interdependence of corn and human beings.

Research cultural beliefs about when to plant and find the scientific basis that supports these beliefs. Have students conduct soil tests and study the climatic conditions to determine if corn would grow well in your area. Find out which American Indian peoples lived in your area before Europeans arrived and determine if they grew corn. What other foods were eaten to maintain a nutritionally balanced diet? What would be the right kind of corn seed to plant in your climatic conditions?

An expansive science unit on corn focuses study on the parts of the plant, genetics, pollination, fertilizers, erosion, intercropping, cross fertilization, hybridization, productivity, and varieties of corn. To keep the science unit within a cultural context study the ecological balance of hill planting relating that balance to the Three Sisters stories. Science experiments compare planting times, days of sunlight, amount of precipitation, charting the results on different styles of charts, such as bar graphs and pie charts, as ways to present information. Collect varieties of corn from many parts of the hemisphere (seed catalogs are a good source) and have students analyze and classify characteristics, and climatic conditions required for each variety.

Making tools from nature's resources provides a way to study indigenous technology. Projects include explaining how corn soup was made using pre-contact tools made from nature's elements including clay, water, trees, ashes, corn, and beans. Evidence of selective adoption in the post-contact period would include use of cast iron kettles and metal corn grinders. Have students compare and contrast making tools from nature's elements and metal or machine-made tools. Are the latter always the best tools for the job? Research technology cross-culturally by finding examples of corn pounders and grinding stones.

Corn was a way to measure time, such as by indicating that a child was born in the season of planting. Have students think of events in their lives that could be described using the agricultural and seasonal cycles. Perhaps they have moved recently; what season was it? Maybe they visit their grandparents when the corn is knee high, when would that be?

Corn was also used to take a census. A Seneca man estimated the 1687 population of the ancient village of Ganondagan by saying one kernel of corn for each person would fill five one-gallon containers. Word problems can be developed to have students practice estimating skills and tabulation.

Conduct a study on the nutritional value of corn and the process of washing the corn with ashes to release the amino acids for better absorption in the body. Study the complete protein that is created by combining corn and beans. Find Haudenosaunee foods that included corn and beans and prepare them in the classroom. Explain that the Three Sisters support each other in the stories, the hill planting, and in foods prepared from them.

Making corn husk dolls and matching them to stories creates an activity that goes beyond the artifact approach. Design and make corn husk dolls to be characters in the corn stories and enact corn stories that express an aspect of the Haudenosaunee cultural paradigm. Examine the many practical uses of corn husks to make mats, pillows, mattress filling, salt containers, and shoes (see Parker 1910).

Bring the course of study to the modern day by having students research both the continued use of corn by the Haudenosaunee people and the commercial uses of corn. For example, how does corn become corn flakes? How is corn syrup processed? Go to a grocery store and read labels to find products made from corn. How is the hominy found in the grocery store processed? How is it different from Haudenosaunee methods?

Interaction of Cultures—Columbus took corn to Europe and it was dispersed throughout Asia and Africa. Have students research and identify the places around the world where corn is grown today. A study of the colonial period can reveal how corn impacted the early settlers, from use as payment for taxes and debts to being their major crop and source of survival. Looking at corn this way enables educators to go well beyond the Squanto story.

Examine how the loss of lands due to wars, treaties, and government policy affected the raising of corn among the Haudenosaunee (Dennonville, Sullivan, Canandaigua Treaty, reservations, Allotment Act, boarding schools, and relocation programs). Create a time line of these important Haudenosaunee historical events and add American historical events to provide a sense of how these time lines coincide. Have students write a research report on the ways culture, way of life, thematic focus, and history are interconnected within a specific cultural area.

Dynamic—Compare the Ernest Smith paintings with historic and contemporary photographs of Haudenosaunee life to identify examples of items selectively adopted and adapted to fit into Haudenosaunee life. For example, cloth was incorporated but clothes were made in the same style as buckskin clothing, or beads were used to replace porcupine quills in decoration, cast iron kettles replaced pottery. What did Americans selectively adopt and adapt from the Haudenosaunee? Corn is certainly a prime example as it was selectively adopted as a food for humans and animals, but the ceremonial aspects were not adopted.

Contemporary Cultural Continuity/Living History—Interviewing Haudenosaunee Elders about the role of world view and corn in their lives provides the basis for a living history project. Write up the interviews as biographies, using pictures if possible. If the Elders are willing, use a video camera, edit, and produce a finished production. Have students write a creative story on five hundred years in the life of a cornfield from the viewpoint of the corn, or research and write an article on "Corn in My Kitchen." Analyze contemporary Haudenosaunee art to identify significant cultural symbols that continue to appear. Use current Indian newspapers to analyze current events.

The interdisciplinary examples provided above are a way to develop creative activities to utilize in developing a culture-based curriculum. It is important when developing activities that you ask yourself if this particular activity can be related to the culturally specific paradigm. If not, the activity could trivialize the importance of the culture. The purpose of culture-based curriculum is to broaden, to expand, students' knowledge and understanding of diverse cultures by moving away from trivialization, artifacts, and stereotypes.

Evaluation

The best form of evaluation consists of a portfolio of student work, including research projects, oral presentations, art work, maps, science experiments, retold cultural stories, and a variety of individual or small group projects. Having students read a treaty, analyze the content and context, and present the treaty to the class through role-play places the treaty and the culture within the historical context and provides a way to grasp the interaction between cultures. Using Figures 4.1 and 4.2 as models, a culturally specific paradigm chart can be constructed for the unique culture students are researching. The outer circle would be the culturally specific paradigm; at the center of the circle would be the thematic focus. At the end of this course of study students will be able to explain, orally or in writing or art, how the thematic focus relates to each element in the culturally specific paradigm. Other innovative ways to evaluate a course of study on a specific culture, or a cross-cultural study, include evaluating individual or small group projects, student interviews of Elders, oral presentations, plays, creating or analyzing works of art, and role-playing or critiquing current issues based on multiple perspectives.

The Case Study

The research methodology selected for this case study is naturalistic inquiry with a focus on identifying "connecting themes" that "describe and explain the system."[20] Using this approach to learn about a culture leads one to see themes from within the culture that provide the culturally specific paradigm and thematic focus. A network or pattern of interconnecting cultural themes emerges to provide a basis for understanding how each theme is a significant component of the holistic culture. This study does not impose meanings on the culture, but rather *seeks to grasp meanings emerging from the culture*.

This case study examines the Haudenosaunee, "The People of the Longhouse," otherwise known as the Iroquois or Six Nations of New York State. The Mohawk, Oneida, Onondaga, Cayuga, Seneca, and Tuscarora are the member Nations of the Haudenosaunee. This study examines the world view of the Haudenosaunee and the role of white flint corn from their creation story to the present day. Corn was identified as a thematic focus for each of the five categories of the cultural research framework:

Chapter 5: The Thanksgiving Address: An Expression of Haudenosaunee World View

Chapter 6: Corn as a Cultural Center of the Haudenosaunee Way of Life

Chapter 7: The Interaction of Corn and Cultures

Chapter 8: Dynamic Aspects of Haudenosaunee Culture

Chapter 9: The Contemporary Role of Corn in Haudenosaunee Culture

Research Methods

Three data sources provide cultural information necessary for developing culture-based curriculum for this case study:

1. Document review of historical documents beginning in the 1500s.

2. Paintings of Ernest Smith, Tonawanda Seneca, (1907–75), 240 paintings completed between 1935–41.

3. Interviews with Elders during 1991–92.

Triangulation, that is, the analysis of information from three sources, provides a comprehensive base of information. These sources provided multiple perspectives to find characteristic patterns and to delve into the meanings of these patterns for the Haudenosaunee. Comparing data over a broad time span, from oral traditions to recorded historical documents, reveals the complexity and continuity of Haudenosaunee culture. Comparing oral traditions with accounts of early explorers and ethnographic accounts provides multiple perspectives on beliefs and events.

Document Review Criteria

Historical documents are an integral source of data for this study because they provide evidence on the Haudenosaunee way of life from the first interactions between explorers and Haudenosaunee people. These accounts serve to confirm the oral traditions, *but* are approached with caution and a discerning eye for the bias. If one can read around the stereotypical words and dismiss the hierarchical scale, the documentation provides invaluable information on events, way of life, and interaction between cultures. Ogawa and Malen define document reviews as multivocal literatures that "reflect different purposes, perspectives, and information bases."[21]

Paintings criteria

The paintings of Ernest Smith, completed between 1935–41, provide visual images of ancient beliefs, stories, lifestyle, historical events, and adaptations. The paintings were analyzed to find patterns, such as the pre–European contact era indicated by people wearing buckskin clothing and using pottery, contrasted with the post-contact era, which can be identified by cloth clothing and iron kettles. Selective adoption and cultural adaptation of items such as cloth clothing not only indicates post-contact, but also illustrates ceremonial clothing worn in contemporary times, thus providing evidence of cultural continuity.

The paintings were analyzed around these culturally specific themes: 1) Thanksgiving Address, 2) corn as significant component of the culture for the thematic focus, 3) the Creation story, 4) the Great Law of Peace, 5) Code of Handsome Lake, 6) Historical interaction, 7) ceremonies, and 8) cultural stories.

Each of these paintings presents cultural information on the Haudenosaunee that would be extremely useful in a classroom setting to provide visual imagery of the Haudenosaunee world view, lifestyle, and interaction of cultures. Since these paintings were completed in 1941, there is the danger of the "living fossils" stereotypes being perpetuated, therefore it is imperative that contemporary photographs and artwork be included in any curriculum utilizing these paintings.

Interviews with Elders criteria

Interviews with contemporary Elders provide evidence of the continuity of the complex structure of the Haudenosaunee way of life. Elders are the storehouse of cultural beliefs and the way of life of the Haudenosaunee. A basic Haudenosaunee cultural value requires younger people to honor and respect their Elders' knowledge. Contemporary Elders have witnessed enormous changes in lifestyle, yet because they have maintained the continuity of the Haudenosaunee world view, they can provide invaluable information around the thematic focus on corn. From these three sources the complex cultural patterns of the Haudenosaunee can be identified from a time period long before written history to the contemporary world. The study is hermeneutical, that is seeking meaning, because these sources provide insight into the cultural meanings of Haudenosaunee perspectives and are studied within the context of the Haudenosaunee way of life.

Thematic Focus

The identification and use of a cultural thematic focus provides a way to demonstrate multiple perspectives around a specific theme. Because the Three Sisters, (corn, beans, and squash) hold such a vital role in the Haudenosaunee way of life, corn has been selected as the thematic focus. Corn is present in the Thanksgiving Address, the three major epic narratives, the ceremonial cycle, and the oral traditions, and formed the economic base of Haudenosaunee society. Corn in contemporary times continues to be used for ceremonies, but has changed from the daily staple to a food for special occasions. The world view continues to be expressed through the ceremonial cycle that revolves around the agricultural cycle.

Corn is a central feature not only of Haudenosaunee culture, but of many Indigenous cultures of this hemisphere. Corn is indigenous to the western hemisphere. Corn is truly a "gift to the world" from the peoples of North, Central, and South America. After the encounter of the two Old Worlds, Columbus began dispersing corn to Asia, Africa, and Europe. During the interaction of cultures in early colonial days, corn was a pivotal economic factor and affected the economy of this country. Corn became an important crop for the United States, and ultimately affected the world economy.

Corn provides a medium to take abstract, or spiritual concepts, such as world view and way of life, and make them tangible for curriculum development and instructional materials. The Ernest Smith paintings provide a visual medium with which to teach about culture. Following the thematic focus of corn from cosmology and world view through structuring the way of life, to interaction of cultures, and to the contemporary world, provides a holistic view of Haudenosaunee culture.

Conclusions

Because culture-based curriculum maintains the perspective that all cultures have value and dignity, it provides a framework to take the societal goal of respecting multiple cultures into educational practice. Evolutionary scales that place diverse cultures on a hierarchical scale undergird the stereotypes, omissions, and inaccuracies in textbooks and media. Culture-based curriculum serves as a model to produce citizens who respect the cultural diversity of this nation. It also provides a way to transform the mainstream curriculum to include diverse perspectives. This model provides a way for instruction to move beyond artifacts, contributions, additive units, and stereotypes.

The components of cultural research framework are: 1) understanding world view; 2) examining how world view structures a people's way of life; 3) studying how cultures interact; 4) understanding the dynamic nature of cultures; and 5) gaining knowledge of contemporary continuity of the culture.

The culturally specific paradigm consists of the world view and how that world view structures a way of life for a specific culture. Continuity of that culturally specific paradigm explains how the people maintain their culture, and why people from a specific culture initiate and respond to current issues in a certain way.

The thematic focus emerging from the culturally specific paradigm forms the foundation for developing an interdisciplinary culture-based curriculum. The process of creating a culture-based curriculum must begin within the culture and emerge from the knowledgeable culture-bearing Elders working with educators. Culture-based curriculum provides a framework for curriculum design that avoids stereotypes by presenting cultures as holistic entities comprised of interrelated cultural components. In the case study, corn emerged as a vital element of the Haudenosaunee culture on spiritual, philosophical, political, sociological, and economic levels, thus corn becomes an appropriate thematic focus.

The culturally specific paradigm emerging from the case study can be applied to diverse cultures by using the All Cultures Chart (Figure 4.1) to identify the interrelated aspects of a culture. In the case study, historical documentation, the Ernest Smith paintings, and interviews with contemporary Haudenosaunee elders provided evidence of the world view and the role of corn into the contemporary time period thus providing information for culture-based curriculum development.

The culture-based curriculum framework allows an integration of culturally based materials and an interdisciplinary approach across the curriculum. The underlying assumptions are valuing cultural diversity, presenting cultures in a holistic manner, and studying the living history contained in the contemporary culture. Developing curricula from this model requires a team approach that includes Elders from the culture, identification of the culturally specific paradigm, selecting a thematic focus to be taught in an interdisciplinary manner, and an analysis of contemporary world view and current issues.

Culture-based curriculum requires one to be of a good mind and have respect for all cultures on Mother Earth.

Chapter 5

The Thanksgiving Address: An Expression of Haudenosaunee World View

The Thanksgiving Address defines and expresses the world view of the Haudenosaunee (Iroquois or Six Nations). The Address can be found in the three major epic narratives in Haudenosaunee culture: the Creation Story, the Great Law of Peace, and the Code of Handsome Lake. Within each of the three epic narratives the Thanksgiving Address is reinforced, expressed, and valued. Together they form the "cultural systems of knowledge"[1] and the "culturally-specific paradigm."[2]

The Thanksgiving Address has its origins in Haudenosaunee Creation. It is reinforced in the Great Law of Peace, which provided a system of government and in the Code of Handsome Lake, which outlines a way to continue the old ways and adapt to the realities of the year 1800. The Thanksgiving Address was spoken at the opening and closing of all ceremonial and governmental gatherings, and this practice continues to the present day.

This chapter provides an overview of the Thanksgiving Address and brief accounts of the three major epic narratives: The Creation, The Great Law of Peace, and The Code of Handsome Lake. Each of these epic narratives is lengthy, extremely detailed, and requires many days to relate in the oral traditional ceremonies. The account presented here, therefore, briefly highlights the major events and themes of the narrative.

Thanksgiving Address

There are generally fifteen or sixteen sections in the Thanksgiving Address, which, Corbett Sundown (Seneca) stated, "correspond to an order observable in nature and represents the sequence of creation."[3] The order begins with the people, the earth, and moves upward to the Creator.

1. The People	9. The Wind
2. The Earth, Our Mother	10. The Thunderers, Grandfathers
3. The Plants	11. The Elder Brother Sun
4. The Water	12. The Grandmother Moon
5. The Trees	13. The Stars
6. The Animals	14. The Four Beings
7. The Birds	15. Handsome Lake
8. The Three Sisters (corn, beans, and squash)	16. The Creator, and the ending, thankfulness for anything which was overlooked

Figure 5.1. Components of Thanksgiving Address

Together these components establish the worldview, the Haudenosaunee "conceptualization of the cosmos."[4] Chafe offers a discussion on the use of the English word "thanksgiving."

> The word "thanksgiving" seems no worse a choice than any other and has been used by most previous writers. When confronted with the Seneca words involved, some speakers balk at any attempt to give an English equivalent. Others translate, to some extent according to context, as *"thank, be grateful to or for, rejoice in, bless, greet."* The trouble is that the Seneca concept is broader than that expressed in any simple English term, and covers not only the conventionalized amenities of both thanking and greeting, but also a more *general feeling of happiness over the existence of something or someone."*[5] (emphasis added)

This Address establishes how the Haudenosaunee, as humans, are interconnected with the universe. Spoken at the beginning and closing of all ceremonies and Grand Council (government) gatherings, the Address keeps these principles ever present in peoples minds.

The origin of the Thanksgiving Address is found in the Haudenosaunee creation cosmology.[6] The Good Minded Twin defines the elements of the natural world included in the Address and establishes the attitude of thankfulness. Within this cosmology, the original instructions to the people direct them to be of a mind that includes thankfulness, peace of mind, an understanding of duty and responsibility, love for one another in kinship, and overall happiness. The Good Mind weaves these elements into "an organic whole."[7]

Except for some deviation in the first section, all of the items are treated in accordance with a fixed pattern. Each section opens with a statement asserting that the Creator decided on the existence of the item and gave each item a purpose that benefits mankind. The speaker explains that the item is still present and carrying out its duties and responsibilities. Finally, those present are asked to express thankfulness that the item continues to exist and carry out its duties and responsibilities.[8]

Kinship terms such as mother, sisters, elder brother, grandmother, and grandfather each designate a family relationship. The Three Sisters (corn, beans, and squash) are called "Dio'he'ko" (in Seneca) which means "our sustenance," "our sustainers." At the end of each section the speaker asks the people to be of one mind in acknowledging and being thankful for those items named and discussed in the section. The people respond saying "nyoh" which expresses agreement and affirmation of the speaker's message. Eloquence in presenting the Address is highly praised among the Haudenosaunee.

One version was related to Foster by Enos Williams, (Mohawk) from the Grand River Reserve in Canada (1970). An earlier version from Chief Corbett Sundown (Tonawanda Seneca) (1959) was documented by Chafe. Although the accounts given here are relatively recent, the Thanksgiving Address was documented by Morgan (1850), Hewitt (1899), and as recently as 1990 by Wall. When Morgan published the Thanksgiving Address in the 1850s, he had heard the Address given at Midwinter ceremonies and it was written down by Ely Parker who said this was his grandfather's version, describing it as "the ancient address handed down from generation to generation . . . Sose'ha'wa has delivered it thus for the past twenty five years at Tonawanda."[9]

Corn in the Thanksgiving Address

The section of the Address relating to the Three Sisters (corn, beans, and squash) as described by Seneca elder Corbett Sundown:

And now this is what Our Creator did. It was indeed at this time that he thought, "I shall leave them on the earth, and the people moving about will then take care of themselves. People will put them in the earth, they will mature of their own accord, people will harvest them and be happy." And up to the present time we have indeed seen them. When they emerge from the earth we see them. They bring us content-ment. They come again with the change of the wind. And they strengthen our breath (life). And when the Good Message came we were advised that they too should always be included in the ceremonies, in the Four Rituals. Those who take care of them every day asked, too, that they be sisters. And at that time there arose a relationship between them: we shall say "the Sisters, our sustenance" when we want to refer to them. And it is true: we are content up to the present time, for we see them growing. And give it your thought, that we may do it properly: we now give thanks for the Sisters, our suste-nance. And our minds will continue to be so.[10]

Enos Williams's section on Our Sustenance:

And now we will speak again. When he made people for him-self, we who are moving about on the earth, He deliberated carefully on the matter: It would not be possible for them to manage alone, the people moving about on the earth. And so he left what are called Our Sustenance. And for this he sent seeds. He decided, "This will sustain them." Moreover he decided, "There will be people in the families that will have hard work to do. It will be their occupation during the day, When the wind turns warm, To select garden spots and work the land. And then carefully they will place them underground, our Sustenance. Then people will beg that it be possible that they grow well."

And indeed it has come to pass. For when the wind turned warm, People placed them underground in the gardens they had selected for planting. And then we saw them growing, Our Sustenance.

And people saw new things hanging. It was possible also for the people to be gathered. They have called it the Green Bean Festival. And we thanked him, he who in the sky dwells, Our Creator. Indeed, the ceremony was passed; we did thank him in the manner he prescribed we should always give thanks when people see anything he has given.

And so it came to pass during our last season. And surely it went on, that they were growing, Our Sustenance, For indeed we saw them again. The Sisters, Our Sustenance, came in again. Then the people we are able to be gathered at the great doings. He decided, "There will be a ceremony." And we think that it came to pass, that the ceremony was held during the last season (Green Corn ceremony). It was a pleasure to see how far Our Sustenance had advanced. How lucky we were to see them grow to maturity again. And people were able to store them away. And truly everyone did give thanks (Harvest Festival). So let there now be gratitude, For they still have their minds fixed strongly upon them Our Sustenance.

There are, still, children standing upon the earth. And it is true, also, that they are still running about. And it is upon them that Our Sustenance have strongly fixed their minds. And we have been able to get all our well-being, our happiness, from them. This is why we are grateful.

From Our Sustenance, the Sisters that he left, we have been getting many uses, day and night. They strengthen our bodies, we who move about on his world. So let us be grateful, let us think upon it, for surely he is sending them, Our Sustenance; and carefully we thank him, he who in the sky dwells, Our Creator, and so it will be in our minds.[11]

The above selections make it clear that the interdependence, the interrelationship, the co-existence of humans and plant life is of paramount importance in Haudenosaunee world view. Because the people did not take it for granted that the crops would grow, an attitude of thankfulness and acknowledgement formed a guiding principle of the culture. The cycles of food depended on nature for the sun, wind, rain, and the cycle of agriculture requires humans to plant, weed, harvest, and consume the corn, beans, and squash. This creates a symbiotic relationship that the Thanksgiving Address acknowledges.

The Sections of the Thanksgiving Address

Each section begins with naming the elements the Creator provided, and stating the purpose, duties, responsibilities, and interrelatedness of the element. Throughout the section people are reminded to be grateful and give thanks, and each section ends with "so it will be in our minds." The people respond saying "nyoh" which expresses agreement. (The Enos Williams version as recorded by Foster is the main source for this overview. The Corbett Sundown version as recorded by Chafe is cited to include more information or to provide clarification.)

The People

> And this is what Our Creator did: he decided, "The people moving about on the earth will simply come to express their gratitude." And that is the obligation of those of us who are gathered: that we continue to be grateful.[12]

> The Sky Dwellers said people should be thankful and give the Thanksgiving Address when they meet each other, when they gather, when they conduct a ceremony. The people were instructed to greet each other, being grateful for their health and happiness, and be content. If any are sick, that is in the Creator's hands. The speaker asks for health for everyone.

The Earth

The Creator made a world "below the Sky World" where people would live. The people are related to the earth, "Our Mother." All the things he created and left on this earth contribute to the people's happiness.[13]

> And it is so still.
> It is possible that it comes from the earth,
> the happiness we are obtaining.
> For all this, therefore, let there be gratitude.[14]

Bodies of Water

The Creator created large bodies of water, streams, and springs to strengthen the earth. Sundown names brooks, flowing rivers, ponds, and lakes.[15] Water will refresh the people; water will bring happiness.

Grasses

He created grasses of many varieties which would grow anew in the spring and bring pleasure to the people, "including children."

The spring breeze which carries the "scent of flowering things" will bring people happiness.[16] Some of these plants will be medicines to heal people, and good health brings happiness. Some plants are utilized as food. Each plant has a name and a purpose.

Hanging Fruit

"He who in the sky dwells, our Creator" made the hanging fruits. The fruits grow in sequence, from just above the grasses, to small bushes, and fruit trees. They also ripen in that natural sequence, meaning strawberries ripen first, then berries (raspberry, blackberry), then apples.

The people are to gather to give thanks, in the Strawberry Ceremony, as this is the first of these fruits to arrive in the spring. The fruits serve to make people happy and are medicines. Sundown includes the Strawberry Ceremony, in which the people make a beverage of strawberries and water (and maple syrup), with the plants section.[17]

Trees

The Creator made trees, forests, with many uses for people. Each has a name. They are used as medicines and to keep people warm (houses and heat) and will bring happiness to the people.

The leader of the trees is the maple. The sap will be gathered and boiled for maple syrup. The people will gather for the Maple Ceremony, to give thanks, to be grateful.

Wild Animals

He created small animals, of many kinds, each with its own name. The leader of the animals is the deer, which the men are to hunt to feed the people. The people will make use of the deer in many ways. The Creator made fish, the underwater creatures.

Birds

Our Creator decided that the trees in the forest would be the home of many different sizes and kinds of birds. When the wind turns warm in the spring, the people will be happy to hear certain birds singing. "And it will lift the minds of all who remain when the small birds return."[18]

> And the minds of the children running about
> will be strengthened.
> Everyone will be happy, the people moving
> about on the earth.
> And he is sending our happiness, our contentment,
> those of us moving about on the earth.[19]

Our Sustenance

Our Creator thought carefully about the people and knew they would need food so he gave the people "Our Sustenance," the Three Sisters, corn, beans, and squash. He gave the seeds and said "the families" will plant "a garden and work the land." He explained that the seeds were to be planted when the wind turns warm and the people will then gather for the Seed Planting Ceremony. The Green Bean Festival is to be held when the beans are ripe. The speaker then relates that these things did happen in the previous year and the ceremonies were held.

> It was a pleasure to see how far Our Sustenance had advanced.
> How lucky we were to see them grow to maturity again.
> And people were able to store them away
> And truly everyone did give thanks. (Harvest Festival)

The speaker refers to the children and says Our Sustenance has strengthened their minds. (A footnote says children are nearer to heaven and there will some day be no children and the food will also be gone.)[20] The Sisters' purpose is to strengthen our bodies.

Sundown says that Handsome Lake reminded the people of the Creator's instructions that the Three Sisters "should always be included in the ceremonies, in the Four Rituals."[21]

The Thunderers

The Address now moves upward from the earth. Williams explains that the Creator "appointed a series of helpers to have certain responsibilities."[22] The first helpers come from the west and are called "Our grandparents, the Thunderers."[23] Their responsibilities are to make fresh water for the bodies of water, wash the earth, and provide water for the gardens.

It is also the Thunderers' responsibility to keep the monsters, which he did not create, below the earth, because the monsters are destructive. (This refers to the Evil Twin and his creations.)

The Sun

The Creator decided there will be day and night. The "Sun, Our Elder Brother," another of his helpers, will have the duty to bring the daylight and give warmth.[24] The sun will move across the sky and "always go in a certain direction."[25] The sun assists Our Sustenance in their growth.

> He keeps the wind at a certain velocity, for the happiness of those of us who are moving about.[26]

The Moon and Stars

The next helper is "another orb in the sky. People will say, 'our grandmother, the moon.'" The Moon is female and her responsibilities are to bring the night, to have phases in her cycle, to bring children, to help raise the children, to control the dew for the plants, to look after the Three Sisters. The Stars work together with the moon.

Sundown has a separate section for the Stars. He says the stars will have names, and will be used to establish directions if one becomes lost. He says the Stars provide the moisture during the night.[27]

The Four Beings

The Four Beings are the Creator's helpers. "We call them the Sky Dwellers."[28] Sundown calls them "Our Protectors" because people have unforeseen accidents.[29] Their responsibility is to guide and help the people because people are sometimes unhappy. They are to straighten out the minds of people and this makes happiness possible.

The Wind

He created winds that have a "certain velocity" for the "happiness of the people." That is, the wind provides air for people to breathe.

> There is a certain place where the wind originates,
> A "veil" people will call it.
> . . . And it will have just the right velocity
> for people's happiness.[30]

Sundown includes the caution that this wind can become too strong and they have seen the wind destroy homes.[31]

Handsome Lake

The preacher explains that people "stumbled . . . when they should have seen their happiness." Sundown explains: "It seemed that nowhere was there any longer any guidance for the minds of those who moved about."[32] Because people made mistakes, the Creator sent his helpers to Handsome Lake and gave him the Good Message to tell the people. "This will be the solution for people in the future."[33] The people are to follow the Good Message.

The Creator

Our Creator lives in the Sky World and listens "intently to us." At this point, the speaker says the people are thankful for any items that may have been left out or overlooked.

> Let there be gratitude day and night
> for the happiness he has given us.
> . . . he gave us the means to set right that which
> divides us. And we may still have our happiness.[34]

Sundown reiterates the original instructions:

> They will simply continue to have gratitude for everything
> they see that I created on the earth, and for everything
> they see that is growing.
> The people moving about on the earth will have love;
> they will simply be thankful.[35]

The speaker tells the people that he has completed the Address as well as it is possible for him to do. He encourages the people to follow the Address and in so doing they will find happiness.

Summary

The world view expressed in the Thanksgiving Address recognizes the complex interdependence and interrelationship of the earth, nature, and human beings. It specifies the duties assigned to each of the elements of the natural world and the duties of human beings regarding each of those elements. It expresses an attitude of appreciation and responsibility in an interconnected whole. Equality is expressed in the interdependence of all that was created. There is no separation between human beings and the natural world; all are equal and interrelated by kinship terms.

The next sections will relate brief versions of the three epic narratives; The Creation Cosmology, the Great Law of Peace, and the Code of Handsome Lake, which contain the essence of Haudenosaunee beliefs.

Creation Cosmology

The first epic narrative is the Creation story. There are many versions of the Creation story which differ because in the oral tradition some speakers provide a long, detailed narrative of the Sky World, while others place emphasis on the section of the narrative about twin forces here on earth. Although the emphasis or details change depending on the speaker, the basic story remains the same. The Onondaga version used in this study was related by John Buck from Grand River to J.N.B. Hewitt in 1889 and revised by his son Joshua Buck in 1897. Hewitt also recorded Seneca and Mohawk versions in 1896–7. John Arthur Gibson told an intricately detailed version to Hewitt in 1900. Each version is published in the language, with literal translations, and an English version. The English version uses terminology that is cumbersome in many respects. It uses archaic language such as "verily," "ye," "thee," and "thou," which one must suspect was used in the early 1900s to enable Hewitt to publish this manuscript for the general public. Hewitt tends to use terms from Greek mythology, such as "Zephyrs," to provide English speakers with a way to connect to mythological concepts.

The Creation

Haudenosaunee cosmology begins in the Sky World, a place just beyond the limits of the visible sky. In the Sky World village lives a young girl and her father. They do not know of death in the Sky World so when her father dies it is an unusual occurrence. His body is placed in a burial case and placed up high, some say in a tree. As the young girl grows up she talks to the body of her deceased father. She is told to travel to the village of He-Holds-the-Earth, keeper of the celestial tree. This tree in the center of the Sky World is full of blossoms that provide light and fruits (food) for the Sky World. She is to marry He-Holds-the-Earth. Her mother makes o'ha'gwa, corn bread, which is packed in a basket carried with a forehead strap. She is warned not to talk to anyone on the way and when she arrives she is to announce to him that they will marry.

She leaves the next morning, returning once because she thinks she is lost. She sets out again and does not talk to Aurora Borealis and the Fire Dragon of the Storm, who try to distract her. She arrives at He-Holds-the-Earth's lodge and rests for the night.

In the morning he gives her a string of white corn and tells her to make mush. First she must soak the corn. He instructs her to put water on the fire to boil and when it is boiling to add the corn and when it is done to remove the pot and wash the corn. Then, the corn must be pounded into meal and the mush made. He tells her to stir the mush constantly and she is to undress. As it cooks, the hot mush splatters on her, but she doesn't complain. He-Holds-the-Earth, the chief, announces the mush is done and removes the pot from the fire. He takes her basket of corn bread and announces they will marry. "So it seems, thou wert able to do it. Hitherto, no one from anywhere has been able to do it."[36] He calls in two large white dogs with tongues like rough bark, who lick the mush from her body. He then uses sunflower oil to soothe her body.

She stays two nights, and it is said that they sleep with the soles of their feet touching. On the third day she returns to her village with a pack basket of dried meat, a gift from the Chief. He had instructed her to tell the people to remove the roofs of their lodges, and had said he would send corn because "that is what man-beings will next in time live upon."[37] She does so, and that night it rains corn into the lodges. The chief of her village instructs the people, "Do ye severally repair your lodges. And, moreover, ye must care for it and greatly esteem it."[38]

She returns to the village of He-Holds-the-Earth. It is said she catches his breath and becomes pregnant. The Chief is very jealous and becomes gravely ill. He has a dream and the entire village comes to guess his dream. Those who try to guess his dream are the deer, buck, spotted fawn, bear, beaver, wind, daylight, night, star, water-of-springs, corn, bean, squash, sunflower, fire dragon, rattle, red meteor, spring wind, turtle, otter, wolf, duck, fresh water, and yellowhammer. Aurora Borelais, the fire dragon or blue panther, guesses his dream.

The dream requires the uprooting of the Celestial Tree. The Chief asks to be laid next to the hole. Some say the child is already born and she ties the child on her back with her shawl. The Chief gives her three ears of corn, dried meat, and wood which she puts into her undergarments. She is afraid as she sits with her legs over the edge of the hole and grasps the edges tightly. As she looks into the dark hole, He-Holds-the-Earth pushes her. She grasps the edge of the Sky World as she falls and she and the child become one again. The Chief is now well and the Celestial Tree is replaced.

As she is falling the Fire Dragon of the pure white body meets her halfway and explains he had made the Chief jealous by telling him the child she carried is not his. He gives her corn, mortar and pestle, a small pot, and a bone saying these will sustain her in the world to which she is falling.[39]

As Sky Woman is falling through the darkness into this world of water, the creatures of this world look up and see her. They hold council and decide to assist her. The loons fly up and gently catch her and bring her down placing her on the back of a turtle; thus this world is known as Turtle Island. The creatures decide they must dive deep into the water to get earth from the bottom of the lake. Many animals try, but their bodies float to the surface. Finally, the muskrat's body floats to the surface, but in his paws he clutches some earth. They place this earth on the back of the Great Turtle and the earth begins to grow.

The next morning when Sky Woman awakes there is a deer lying there for her to eat. She makes a fire from the wood which He-Holds-the-Earth had given her. After three days her daughter is born. The child grows rapidly and becomes pregnant with twins, some say by the West Wind.

The daughter hears the twins arguing over how they will be born. The Good Twin is born in the normal manner, but the Evil Twin (Flint) bursts forth through the mother's armpit, which kills her. It is said that from her grave grow the corn, beans, and squash, the sacred tobacco, and wild potatoes. Others say that the father of the twins, the West Wind gives the Good Twin the seeds of corn, beans, squash and tobacco.[40]

The Good Minded Twin creates plants, animals, rivers, trees, berries, birds, and makes human beings from clay. He is recreating the Sky World on this earth. The Evil Mind (Flint) creates poisonous creatures, he puts thorns on the berry bushes, brings disease and hardship; he seeks to undo his brother's work in his attempts to gain control over this world. The Good Twin and the Grandmother play the peach stone game in their struggle over who will control this earth.

Although the forces of creation and destruction are ever in opposition, eventually through battles, including a lacrosse game, the Good Mind wins control over this world. The Good Mind cautions the people, explaining that there will always be an eternal struggle, that people will always be of two minds—good and evil.

In the Gibson version the Good Mind returns to the Sky World on the path of the Milky Way. Four times he returns to this world because the people have gone astray and are in need of more instruction. During these return visits the Good Mind establishes the Thanksgiving Address, the Four Rituals (Feather Dance, Thanksgiving or Drum Dance, Personal Chants, Peach Stone Game), the pattern of communal agriculture relating to the Three Sisters (corn, beans, and squash), explains death, and creates the clan system of families.[41]

The Great Law of Peace

The Great Law of Peace is the epic narrative that established the form of government based on a consensus decision-making process, a confederacy of the five nations. The Peacemaker was a Huron, who traveled to the Haudenosaunee lands to bring the message of peace.

At this time, the people had forgotten the old ways; thus, blood feuds, revenge, even cannibalism had become the prevailing human behavior. The Peacemaker and Aywentha (Hiawatha, Haiyo'wentha) undergo many ordeals in their journey to bring the message of peace, and the understanding that human beings are rational beings capable of negotiating to solve differences. Aywentha had lost his daughters to unseemly deaths and cannot be comforted. The Peacemaker condoles him by providing a method of comforting those who have lost someone. In this way he provides a healing way to deal with death which ends the blood feuds.

In the journey Jigonsaseh is the first woman to accept the Great Peace. Because of her acceptance, women have an honored role in government. They select the fifty chiefs and have the power of impeachment.

The last to accept the message of peace is Tadodaho, the evil wizard of the Onondaga. He is said to be a cannibal; his body is twisted and snakes grow from his head. All the other nations had accepted the message of peace and come together, with Jigonsaseh, to approach Tadodaho. They sing a song, comb the snakes from his hair, untwist his body, and he accepts the great peace. The Peacemaker and Hiawatha (Aywentha) bring a system of government, with rules and protocol, to establish peace and a method of resolving differences.[42] The meetings of this government are called Grand Council and continue to be held in Onondaga, (just south of Syracuse, N.Y.) to the present day.

The Peacemaker instructs that the Grand Council meetings be opened and closed with the Thanksgiving Address spoken by an Onondaga holding the strings of white wampum, which represent peace.[43] The 1885 Newhouse version of the Great Law of Peace instructs the Chiefs to give the Thanksgiving Address and includes the strings of white wampum, thus confirming the oral tradition.[44] Thus, the people are reminded of the elements that comprise the Haudenosaunee world and are asked, once again, to be thankful, to be of one mind, to be grateful, that these elements continue.

The Code of Handsome Lake

The Code of Handsome Lake (the Gaiwiio, the Good Message) contains instructions, directions, a plan to enable the Haudenosaunee to maintain the values of the old ways while adapting to living in the modern world of 1800. Handsome Lake was a Seneca born in 1735 at Conawagus village on the Genesee River. In his lifetime he saw the Seneca removed from their ancestral homeland in the beautiful Genesee valley and resettled on reservations at Buffalo Creek, (current day Buffalo, N.Y.), Tonawanda, Alleghany, and Cattaraugus. Parker describes the conditions of the time period:

> The encroachment of civilization had demoralized the old order of things . . . frauds . . . loss of land . . . poverty. . . . [T]he crushing blow of Sullivan's campaign was yet felt and the wounds then inflicted were fresh . . . poverty . . . defeat . . . loss of ancestral homes . . . broken promises . . . hostility of white settlers . . . despair . . . hopeless. . . . [T]he greedy eyes of their conquerors fastened on the few acres that remain to them. It was little wonder that the Indian sought forgetfulness in the trader's rum.[45]

It was into this time of drunkenness, of despair, that Handsome Lake received a vision from the Creator in 1799. Handsome Lake had been given to drinking most of his lifetime. He was ill for four years and appeared to have died, but he was in a coma-like state and had a vision in which he was given the Great Good Message. The purpose of this message was to maintain the most important, the vital, customs of the past, and provide a way to cope with and adapt to the situation of the 1799. The message has 130 sections and is taught over the course of three days each fall in Haudenosaunee longhouses. I have grouped the 130 sections into six major areas: 1) ceremonial renewal, 2) Three Sisters, 3) unacceptable behavior, 4) acceptable behavior, 5) the two paths/a journey, and 6) sections interwoven with historical people, places, events.

Ceremonial Renewal

In these sections Handsome Lake instructs the people to continue the Four Rituals, (Feather Dance, Thanksgiving or Drum Dance, Personal Chants, and Peace Stone Game) in the ceremonies. "The Creator has ordered that on certain times and occasions there should be thanksgiving ceremonies." He says the wild animals will become extinct and the people are to use "cattle and swine for feast food at the thanksgiving."[46] In two sections he directs the people to give the Thanksgiving Address in the ceremonies, saying,

> It is said that when these rites are performed one person is to be selected to offer thanks to the Creator. Now when thanks are rendered begin with the things upon the ground and thank upward to the things in the new world above.[47]

There are specific sections on the Midwinter ceremony, Planting ceremony, and Strawberry ceremony. These sections continue the world view of the Haudenosaunee by reinforcing, that is reestablishing, the vital importance of the ceremonies to Haudenosaunee culture.

The Three Sisters

> When the leaf of the dogwood is the size of a squirrel's ear, the planting season has come. Before the dawn of the first day of the planting a virgin girl is sent to the fields where she scatters a few grains of corn to the earth as she invokes the assistance of the spirit of the corn for the harvest.[48]

Handsome Lake instructs the people to continue the custom of giving a thanksgiving at planting time. He mentions that the "Dio'he'ko (the corn, bean, and squash spirits), have secret medicine in which the seeds should be soaked before planting. He says the medicines grow on the "flat lands near streams."[49] The ceremonies and agricultural economy based on corn are essential aspects of Haudenosaunee culture which the people are instructed to maintain.

Unacceptable Behavior

One of the main tenets of the Good Message is that the Haudenosaunee should stop drinking. This message is repeated in several sections of the message. The punishment for those people who waste their life drinking is quite dire. The people are instructed to stop listening to gossip or liars, not be to vain or boastful, not to steal food from others but to ask for food, not to strike each other and to stop witchcraft and abortions. The message says to avoid fiddle playing, which represents modern music, because where this music is played there is usually drinking. Playing cards or gambling receives the same restrictions.

Acceptable Behavior

Handsome Lake instructs the people to attend ceremonies with a thankful heart. He emphasizes raising a good family who treat each other well. He warns not to strike a child, but to discipline by giving children three warnings and throwing water in their face, but if they don't behave after the third warning to throw them into the water (stream). He cautions adults to listen to children for they have good minds. People are always to be generous with food and feed everyone who enters their home. They are to help one another by organizing "work bees" as a community endeavor to assist each other with the work.

   One section deals with the major adjustment from having vast areas of land in which to hunt, fish, gather, and raise agricultural crops to the much smaller reservation areas after colonial contact, providing these instructions: 1) cultivate the ground and grow food, 2) build a modest house, and 3) raise horses and cattle for food.

   On the topic of education, meaning non-Haudenosaunee education, Handsome Lake gives the following instructions:

> This concerns education. It is concerning studying in English schools, Now let the Council appoint twelve people to study, two from each nation of the six. So many white people are about you that you must study to know their ways.[50]

The message emphasizes raising a healthy loving family, continuing to work together as a community, and continuing to plant corn, but also beginning to cultivate the land and raise domesticated animals. Their form of agriculture will now change to farming.

Two Paths/A Journey

On the third day of the recitation of the Code of Handsome Lake, he tells of his journey on the "Great Sky-Road." The Sky Road is the Milky Way, the path to the Creator's land in the Sky World.

The messengers show Handsome Lake two paths, one that is wide and has many people on it which leads to the "house of the punishers," and the other is a narrow path which goes to the "lands of Our Creator." He takes Handsome Lake to see various places in the punishers' land where people are punished for wrongdoing in this world. The place is described as a hot place where you hear "mournful cries." The wrongdoings include stinginess, gluttony, the selling of Indian land, drinking, witchcraft, wife beating, quarrelling, immorality, playing violin, and cards. The people in these places did not repent their misbehavior.

There is a fascinating encounter with Jesus on this journey, who says:

> [T]hey slew me because of their independence and unbelief. . . .
> Now let me ask how your people receive your teachings.
>
> He answered, "It is my opinion that half my people are inclined to believe in me."
>
> Then answered, he, "You are more successful than I for some believe in you but none in me. I am inclined to believe that in the end it will also be so with you. . . . Now tell your people that they will become lost when they follow the ways of the white man."[51]

This exchange between prophets proves to be intriguing because the overall message is to balance the old and new, yet, here we have Jesus saying not to follow the ways of the white man. Does this mean not to follow the white man's religion?

On the Creator's path to the Sky World, Handsome Lake describes a brilliant light and smells the "fragrant odors of the flowers along the road" and sees "delicious looking fruits" and many birds. "The most marvelous and beautiful things were on every hand."[52] In this place he sees his own dog, a white dog, which was sacrificed at the Midwinter ceremony. He meets his daughter and a son. He hears a voice calling the people to perform the Great Feather dance. It is the voice of one of his departed friends who is now singing the Feather dance in the Sky World. Handsome Lake then returns to earth because to enter the lodge prepared for him would mean his death and it is not yet his time to die.

Sections Interwoven with Historical People, Places, and Events

The uniqueness of the Code of Handsome Lake, unlike the Creation and the Great Law of Peace, which happened before contact, appears with the references to historical people, places, and events that date the message after contact. One reference is to the Buffalo Creek reservation, which Handsome Lake describes as "honeycombed and covered with a net."[53] The messengers with him say this reservation will fall. (The Buffalo Creek reservation was lost in the fraudulent treaty of 1838. Some people say the "net" over that land represents the power lines we see today flowing from the New York State Power Authority Plant.)

One of the most intriguing items in the Good Message is a reference to George Washington. Handsome Lake sees a house suspended in the air. On the porch of this house a white man is walking with his dog. The messengers identify the man as George Washington. The messengers tell of the time when the "Thirteen Fires and the King" are in a war which the Thirteen Fires win. The King leaves what should happen to the Iroquois in George Washington's hands. According to the messengers, Washington says, "I shall let them live and go back to the places that are theirs for they are an independent people."[54]

Quite another version of the interactions with George Washington was related by Emily Tallchief, the great-great-granddaughter of Cornplanter, who was a half brother to Handsome Lake. In her account, Washington is determined to punish the Iroquois for their part in the war, "for he did not realize that we were keeping our treaties with the British when we fought." Washington orders the Iroquois to move to the west, but Cornplanter refuses, saying, "We have long lived here and intend to continue in our own territory as

long as we are able to hold it."[55] Washington refuses to listen, and
Cornplanter returns to the people with Washington's position. The
western Indians begin to protect their lands and attack settlers.
Cornplanter becomes the intermediary when Washington decides
an Indian war is a mistake. Cornplanter threatens the western
Indians with a war by the Iroquois and peace is accepted.

It is said that George Washington lives in that house suspended in air, which is not in the Creator's land but is on the path
to the Creator's land. Washington's role in the 1779 destruction of
forty Seneca and Cayuga villages causes one to wonder how he
could possibly be given a place of honor. One cannot judge, but it
is important to know there are many perspectives on George Washington and his actions.

Another historic figure, Red Jacket, a Seneca, is described in
the journey. Red Jacket is seen "carrying loads of dirt and depos-
iting them in a certain spot. He carried the earth in a wheel-
barrow. . . . It is true that his work is laborious and this is the
punishment for he was the one who first gave his consent to the
sale of Indian reservations."[56] This section reveals the horror of the
treaty-making period when millions of acres of land were lost. It is
a dire statement intended to prevent the people from selling any
more land or signing more treaties.

Handsome Lake taught his message for sixteen years. He lived
ten years at the Cornplanter reservation, two years at Cold Spring
longhouse (Alleghany reservation), and four years at Tonawanda.
He knew when he was called to Onondaga that he would die there.
He was reluctant to go to Onondaga, but finally did go, and he died.
A stone monument marks his grave at Onondaga. Parker says, "It
is an odd sight, provoking strange thoughts, to stand at the tomb
of the prophet near the council house and watch each day the
hundreds of automobiles that fly by over the State road."[57]

The question of just how much Christian influence affected
the Code of Handsome Lake has been a subject of great debate
since Handsome Lake first began to reveal his experience. Those
who are staunch supporters of the Great Message will argue there
was no influence and consistently reject Christianity. On the other
side of the debate are those who reject the message saying it is too
Christian and Handsome Lake was just an old drunk. The debate
misses the point of the Great Message, which is continuity of the
old ways and yet adaptation to the modern world. It is in essence
one of the finest messages about keeping the very best of the old
and rejecting the very worst of the new.

How the Code of Handsome Lake came to be written down is
an interesting story in itself. According to Parker, the Haudenosaunee

preachers of the Code of Handsome Lake gathered, six at that time, fifty years prior to his writing to compare their versions. Since Parker published the Code in 1913 this discussion took place around 1863. "Chief John Jacket, a Cattaraugus Seneca, and a man well versed in the lore of his people, was chosen to settle forever the words and the form of the Gai'wiio'. This he did by writing it out in the Seneca language by the method taught by Rev. Asher Wright, the Presbyterian missionary." This original has not survived. In 1903 Chief Cornplanter wrote the Good Message "in the old minute book of the Seneca Lacrosse Club." Parker learned of this and encouraged Cornplanter to finish it and have it preserved.

> The translation was made chiefly by William Bluesky, the native lay preacher of the Baptist church. It was a lesson in religious toleration to see the Christian preacher and the "Instructor of the Gai'wiio'" side by side working over the sections of the code, for beyond a few smiles at certain passages, in which Chief Cornplanter himself shared, Mr. Bluesky never showed but that he reverenced every message and revelation of the four messengers.[58]

Parker published this version in 1913. The current-day speakers who present the oral presentation each fall in the Haudenosaunee longhouses are the major sources of information on the Code of Handsome Lake.

Summary

The Creation story, the Great Law of Peace, and the Thanksgiving Address, pre-date contact (1492) in oral tradition. The antiquity of these epics is stated by the Elders, and is readily apparent because the imagery is consistently of the natural world. All the plants and animals that the Good Mind created are the plants and animals indigenous to the northeastern United States. The Code of Handsome Lake reinforces ancient beliefs such as the Thanksgiving Address and the ceremonial cycle, but it also includes historical places, events, and people. Although attempts have been made to put a date on the origins of the Great Law of Peace, they are conjectural. These cultural beliefs have existed into antiquity and cannot be accurately dated.

A consistent pattern emerges through these three epic narratives. In each narrative there are instructions or messages given or renewed because the people have fallen away from the original instructions of thankfulness and peace of mind. We see this in the Creation story when the Good Twin returns four times to give more instructions because each time he goes away the people follow along for awhile, but then fall away from the message. The Great Law of Peace was provided at a time when the people had fallen so far away that they engaged in blood feuds and cannibalism. The Code of Handsome Lake message was given when the people were demolishing themselves with drinking. In each instance, the people are reminded to renew the old ways by giving the Thanksgiving Address, renewing the ceremonies, and continuing to plant the Three Sisters, as well as take up better ways, such as government in Great Law of Peace and adaptation to the modern realities in the Handsome Lake message.

Another pattern, dualism, emerges in these epic narratives. Dualism, the fact that people would always be of two minds, is presented in the Creation story by the Good Twin and Evil Twin. Dualism is evident in the Great Law where the Peacemaker and Aywentha are teaching peace but Tadodaho resists. In the Code of Handsome Lake, dualism is the maintenance of the old ways while adapting to the current situation.

The theme of continuity is evident in Handsome Lake's message when he reinforces the ceremonies, the Four Rituals, the Thanksgiving Address, the Sky-Road or Milky Way, and the Three Sisters, Dio'he'ko, agricultural systems. We find cultural continuity expressed in each of the epic narratives by the renewal of the Thanksgiving Address and ceremonies. Thus, the Thanksgiving Address establishes the world view, and the ceremonies, which will be discussed in the next chapter, establish the way of life.

Chapter 6

Corn as a Cultural Center of the Haudenosaunee Way of Life

The world view of the Haudenosaunee becomes enacted in ceremonial and daily life, which forms their way of life. When we attempt to learn about another culture we must try to understand that culture within a holistic perspective. Cultures are composed of interrelated and interdependent elements, therefore one item can not be artificially extracted from a culture and studied as representative of the entire culture. Utilizing a cultural element, in this instance corn, as a thematic focus provides a way to present the selected cultural element within its connection to *all parts* of the culture, that is, the holistic cultural context.

Corn is integral to all aspects of the Haudenosaunee way of life. In this chapter corn is examined in: 1) the ceremonial cycle; 2) the oral traditions and stories; and 3) the interconnected cultural and agricultural economic base.

The ceremonies of the Haudenosaunee intertwine nature's cycle (maple, thunder, strawberry) and the agricultural cycle (planting, green bean, green corn, and harvest). In the Creation story, the ceremonies were established by the Good Twin during one of his four return journeys to give people the original instructions.[1] The Three Sisters (corn, beans, and squash) and ceremonies are discussed in the Thanksgiving Address.[2] The ceremonies have been handed down from generation to generation from ancient times.[3]

The Haudenosaunee Ceremonial cycle chart describes the major ceremonies of the Haudenosaunee.[4] The number of days of each ceremony and specific components vary from longhouse to longhouse, that is, they are community specific.

CEREMONY	MONTH	HOW DETERMINED	NUMBER OF DAYS
Midwinter	January/ February	held five days after the new moon	7–10
Maple	March	when the maple sap begins to flow	1
Thunder	April	when the thunder is heard for the first time	1
Planting/Seed	May	determined by the moon, when the dogwood leaf is the size of little finger	1
Strawberry	June	when the wild straw- berries are ripe	1
Green Bean	July	when the green beans are ripe	1
Green Corn	August/ September	when the corn is in the milky stage	2–3
Harvest	September/ October	when the crops have been gathered and stored	4

Figure 6.1. Haudenosaunee Ceremonial Cycle

Each of the ceremonies begins and ends with the Thanksgiving Address, which brings people's minds together as one. The Thanksgiving Address expresses the gratitude and thankfulness of the people for the people, the Mother Earth, grasses, plants, trees, animals, waters, birds, Three Sisters, winds, Grandfather thunderers, Elder Brother sun, our Grandmother moon, stars, four messengers, Handsome Lake and The Creator. Speeches relevant to the specific ceremony are given and the people participate in the Great Feather Dance for the Creator and other dances specified for the ceremony. The ceremony ends with the Thanksgiving Address and the huge cast iron kettle of corn soup, or special corn dishes, is brought into the longhouse. The soup is divided into containers which each family brings for this purpose. The families depart to share the ceremonial foods in their homes, although at some longhouses the food is eaten in the longhouse or in a separate building called the cook house.

In addition to these regular ceremonies there are specific gatherings to tell the Great Law of Peace and Code of Handsome Lake. An elder, who is the holder of this knowledge, teaches the Great Law of Peace once every five years, but it can be more often if requested. The Code of Handsome Lake is heard every fall, usually in September, as the holder of this knowledge travels to the various longhouses to relay the message. Parker mentions special annual ceremonies which include the Sun and Moon dance, each held when someone dreams they should take place. Because other ceremonies are private and of a medicinal, or healing nature, or for the departed, they will not be included in this discussion.

Parker describes a special society of women called the Ton'wisas whose "special duty is to offer thanks to the spirits of the corn, the beans, and the squashes, *Dio'he'ko* (these sustain our lives)." The leader of this society carries "an armful of corn and a cake of corn bread," as she leads the line of women around the kettle of corn soup. In her right hand she has beans, and in her left hand she carries squash seeds. The song is accompanied by shaking a small turtle shell.[5] Parker tells a story from the late seventeenth century in which the entire Ton'wisas society was captured by the Cherokee and taken down the Ohio River. Since that event, two men are allowed to escort this society. Jesse Cornplanter, then a young boy, sketched the "Ceremonial march of the Ton'wisas Company." The leader of the women carries ears of corn in one arm and a tortoise shell rattle in her right hand.[6] At Cattaraugus this ceremonial march is called the Moon Dance and is held during the planting ceremony. The women stand in a circle at one end of the longhouse and sing while the men are singing Adonwah or personal chants at the other end of the longhouse. The Moon Dance continues to take place at Cattaraugus, Tonawanda, and at Grand River in the Onondaga longhouse. Because the vital role of this women's society in offering gratitude and thanksgiving was to provide a plentiful harvest, this society combines ceremony and agriculture, just as the corn, beans, and squash are intertwined.

At the end of each ceremony the huge kettle of corn soup, or corn dishes specific to the ceremony, is brought into the longhouse for distribution among the families. Harrington says, "A stranger

boarding with an Indian family often gets his first notice of a ceremony the night before from the o'no'kwa' (corn soup) served at
breakfast." Mention is made of a string of corn "hung as an offering
outside the 'longhouse' " during some of the ceremonies, but other
references to this have not been located. Jesse Cornplanter sketched,
"Preparing for the Green Corn dance," which shows a woman using
a wooden paddle to stir the corn soup in a large cast iron kettle.
Both men and women are shown seated on benches outside the
longhouse husking and shelling corn.[7]

> *The mythology of the Iroquois is full of allusions to corn, its
> cultivation and uses.*[8]

Oral traditions generally fall into three categories: 1) narratives
that explain natural phenomena; 2) narratives that teach a lesson;
and 3) stories for entertainment, adventure, and amusement. The
oral traditions include themes connected to corn such as: the first
mention of corn in the Sky World; the origins of corn; corn in the
Great Law of Peace; corn and medicine; the intertwining of corn,
beans, and squash; lessons in sharing and appreciation of corn;
Tuscarora corn stories; Handsome Lake and spirit of the corn; and
adventure tales.

Corn appears in the Sky World narrative at four points: when
Sky woman's mother made corn bread for her to take to her intended husband; when Sky Woman made mush for He-Holds-The-
Heavens; when he rained corn into the longhouses of her village as
a gift for marrying her; and when the Fire Dragon or Blue Panther
gave her corn and a mortar and pestle as she fell from the Sky
World to this world.

The origin of corn, beans, and squash in this world was said to
have come from the grave of Sky Woman's daughter after she died
giving birth to the Twins. Corn is believed to have come from "breasts
of the Earth-Mother."[9] Parker cites this oral tradition as being the
generally accepted version of the origin of corn. From the same grave
grew squash from her navel, beans from her feet, and tobacco from
her head. In Jesse Cornplanter's drawing, "From the Body of the Old
Woman Grew Their Food," the tobacco, corn, and squash are the
same, but beans grow from her fingers, and the potatoes are at the
feet.[10] Chief Corbett Sundown confirms that the wild potato, today
known as the jerusalem artichoke, grew from her grave.

> When the grandmother saw the plants springing from the
> grave of her daughter and cared for by Good Mind she was

thankful and said, "By these things we shall hereafter live, and they shall be cooked in pots with fire, and the corn shall be your milk and sustain you. You shall make the corn grow in hills like breasts, for from the corn shall flow our living."[11]

The Evil Twin was jealous of the Good Twin's gift of the Three Sisters and spread a blight on the corn which made it harder to grow and less abundant.[12] Cornplanter's drawing, "The Evil-Minded Pours Ashes on the Corn to Spoil It," illustrates this story, although in the drawing the Grandmother throws ashes on the corn. Several versions have different origins of corn. Cornplanter says the Good-minded Twin received corn, beans, squash, and tobacco seeds from his father the West Wind.[13] Gibson said these items came from the grave of the Good Twin. The Three Sisters appear together and although the versions differ, the sources are consistent with the major characters in the story.

There are stories within stories in the Great Law Of Peace. The Peacemaker and Aywentha travelled from nation to nation to convince the Haudenosaunee nations to accept the Good Message of Peace or the Good News of Peace and Power.

At one point in their travels the Peacemaker warned Haywentha of danger. He said a man was watching him. Haywentha approached an Oneida man sitting beside his fire and asked what he was doing. The man replied, "I am watching the fields of corn to protect them from other nations and also from animals that our children might live from the harvest." Haywentha sent the man back to his village to tell the chief that the Great Message had arrived. The chief asked who this man might be and sent the man back to invite him to the village. Haywentha wouldn't travel to the village until he was invited properly, therefore this story establishes the origins of using wampum as the official form of invitation.

The Peacemaker asked who was the first person in the village to hear the Good Message. The chief said it was the man who was guardian of the corn fields. The Peacemaker asked the guardian of the corn fields how he protected the fields, and he replied that he used bow and arrows which were carried in a quiver on his back. The Peacemaker renamed him Oh-dah-tshe-deh, meaning quiver bearer. The Peacemaker said to him, "Your duty hereafter shall now be to see that your children (instead of the fields) shall live in peace."

The Peacemaker asked the Chief how he dealt with the damaged crops and he answered that he gathered the crops and divided the corn equally among the people. The Peacemaker replied, "You shall now therefore hereafter be called Ka-non-kwe-yo-da," which means "A Row of ears of corn standing upright." "It shall therefore now be your duty to propagate the Good Tidings of Peace and Power so that your children may live in peace."[14]

It was the men's responsibility to guard the cornfields, according to this section of the Great Law. The presence of men is significant because later researchers insist that men did not take part in agriculture except to clear the fields and help with the harvest. In this account a transfer took place from men being part of agriculture to being responsible for keeping peace for the village.

The importance of being kind to animals, birds, and all of nature results in human beings being given the gift of medicine in the story of Red Hand.

Red Hand was a young chief and an excellent hunter. He was well known among the animals because he never "shot a swimming deer or a doe with a fawn." He was kind to animals and always left scraps from his hunting for them to eat. He offered tobacco in a ceremony to ask permission to kill animals for food. "When he had his corn harvested he left open ears in the field for the crows, that they might not steal the corn sprouts at the next planting." He was friends with the animals and they respected him. He was known as the "protector of the birds and beasts."

One day he went on an exploring party to the southwest, where new varieties of corn and beans were found. His group was attacked, killed, and scalped. The animals gathered around him and recognized him as their friend and determined they would obtain his scalp from the enemy and make a medicine to restore his life.[15]

In the Curtin version, each animal gave a piece of itself for the medicine and "experimenting with their medicine they caused a stalk of corn to grow out of the ground without sowing seed. In this stalk there was blood . . . they broke the stalk, and after obtaining blood from it, caused it to disappear. With this medicine is compounded the seed of the squash."[16] This medicine and their singing brought Red Hand back to life and formed the medicine society.

In the Parker version, the animals and birds created a medicine and brought Red Hand back to life giving him songs and dances for the medicine society ceremony. As they sang for him they "shook rattles made of the squashes (gourds)." However, the animals did not tell the ingredients of the medicine because Red Hand was a married man and only a virgin man could receive this knowledge. Later, a young man heard singing and had great adventures in his search to find the source of the song. Finally, after surviving a perilous climb up a mountain he found a single great corn stalk growing out of a flat rock. "Its four roots stretched in the four directions, north, east, south and west." The wonderous song came from the cornstalk. "The corn was a mystically magic plant and life was within it." The song told him to cut a piece of root and take it for medicine. He offered tobacco and then, cutting the root, blood flowed from the cut and the stalk was immediately healed. The corn root was part of the secret of the medicine. He was given songs and the ceremony for healing with this secret medicine and thus began the Charm Holder's medicine society.[17] Jesse Cornplanter and Ernest Smith illustrated the story of Red Hand, showing many animals and birds gathered around Red Hand as they give him the medicine.

Corn, beans, and squash are intertwined in Haudenosaunee mythology as in this story related by Joseph Lyon, Ka-no-wash-yen-ton, an Onondaga, in 1898:

A fine young man lived on a small hill, and being there alone he wished to marry. He had flowing robes, and wore long and nodding plumes, so that he was very beautiful to behold. Every morning and evening he came out of his quiet house, and three times he sang, "Say it, Say it. Some one I will marry," and he thought he cared not at all who it might be. For a long time he kept this up, every morning and night, and still he was a lonesome young man. At last a tall young woman came, with long hair neatly braided behind, as in the Indian style. Her beads shone like drops of dew, and her flowing green mantle was adorned with large golden bells. The young man ceased to sing, and she said, "I am the one for whom you have been looking so long, and I am come to marry you." But he looked at her and said, "No! you are not the one. You wander so much from home, and run over the ground so fast, that I cannot keep by your side. I cannot have you." So the pumpkin maiden went away, and the young man was still alone, but kept on singing morning and night, hoping his bride would come. One day there appeared a slender young woman, of graceful form and fair of face. Her beautiful mantle was spotted here and there with lovely clusters of flowers, and groups of bangles hung upon it. She heard the song and drew near the singer. Then she said she could love dearly one so manly, and would marry him if he would love her in turn. The song ceased; he looked at her and was pleased, and said she was just the one he wished, and for whom he had waited so long. They met with a loving embrace, and ever since the slender bean twines closely around the corn, he supporting her and she cherishing him. Perhaps it might be added that they are not divided in death, for beans make a part of Indian corn bread.[18]

Hewitt recorded the story as part of a much longer story on the acquisition of corn. In his version the Bean Maiden is the singer and the wolf, bear, and deer are rejected because they cannot offer her the appropriate food. Finally, Corn Man is accepted by Bean Maiden:

> I will share with you your fortune or misfortune, whichever it be, wherever mankind shall have charge of your welfare and needs, for my grandmother has appointed me to care for mankind during the time that this earth shall endure. So it shall be that they shall plant us always in one place. So from one place you and I together shall depart when the time during which we shall provide (food) for mankind, as had been appointed for us, shall expire. We must teach them our songs and dances, so that mankind may express their gratitude when they shall gather in their harvests of corn and beans and squashes.[19]

Bean Woman continues to discuss with her husband the corn dances, ceremony, and songs. She describes the ceremony in which the women on both sides of the longhouse would form two lines with the clanmothers leading. One side would sing a song and then the other side with the leaders carrying a small turtle shell rattle. After the song the two lines of women would dance three times around the fire singing the A'konwi'se (the ceremony of the Corn dances.) The song is one of preparing the corn for planting by sprouting the corn seed in a bark bowl, of planting, of "a beautiful Spring season," seeing a "fine field of growing corn," and seeing the drying poles and rack for the harvest on which they see "fine fat strings of corn hanging." It ends, "Now, I am walking along. I am walking along giving thanks to the Life God." Hewitt documented another version of this story, "The Bean Woman," in which she rejects the panther, deer, bear, and wolf before accepting Corn Man as her husband.[20] Parker confirms he had heard this story but the corn is female.

Haudenosaunee stories that teach a lesson appear throughout the oral traditions. "Corn Rains Into Empty Barrels," recorded by Arthur Parker, provides an example of sharing and appreciation:

A long time ago the people were starving because there was nothing left to eat. They lived on a hill and existed on boiled bark. Two brothers lived in this village. One was optimistic, he thought things would improve, the other tortured him with stones and harsh words. The first brother told the people he heard footsteps for the "span of the moon" (a month) but they wouldn't believe him. One day a young woman appeared at his door. She gave him a basket of bread and explained that her mother wanted her to marry him. She came from the south where food was plentiful. They married. She told the people to take off the lids of their corn barrels, go inside the lodge and cover their faces. Soon they heard "a sound like corn falling into their barrels." They were happy now and ate well of the corn. The younger brother wanted meat and threw the corn into the fire. The husband caught fish for the people, but the younger brother said there were no fish. The next day the husband went hunting. While he was gone the younger brother tormented the wife and again threw the corn food into the fire. She was crying when her husband returned and said she would return to her home because "her mother instructed her to return if she were abused." She left, and that night the people heard scraping in the corn bins and discovered the next morning that all the corn was gone. Again, they faced starvation. The husband went after his wife on a long difficult journey. He finally reached a corn field and found his wife and her mother on a "high mound covered with corn plants. His wife showed him her body and it was burned and scarred." She explained this had happened to her when the younger brother threw the corn into the fire. He stayed there with her for a few months before they returned to his village where the people were starving. She told them to take the lids off the corn bins and again it rained corn. "Then the young wife told the people that corn must never be wasted or thrown away for it is food and if destroyed will cause the crops to be poor and the corn to cease to yield."[21]

The same story, "The Corn Goddess," is found in Canada. Another version of this story is found in Curtin and Hewitt but the people are living on a high hill due to a flood. The wife does not return with the husband. He overcomes many obstacles on his journey home with the corn and other seeds.[22]

"The Weeping of the Corn, and Bean, and Squash People," develops the theme of sharing and appreciation.

In this village the people raised corn, beans, and squash for a long time, but their crops began to fail and the people faced starvation. An old woman heard weeping in the fields. She went to the field and found the Corn Spirit weeping because the people had neglected the corn by not hilling up the earth around the corn and hoeing the weeds. The weeping Bean Spirit and Squash Spirit told the same story. The old woman returned to her village and wept. The Chief asked her why she wept and she told about the weeping of the Corn, Bean, and Squash Spirits. He called the people together and they decided to plant and care for the crops properly. For two seasons they planted and cared for the crops, but just before harvest thieves took their crops. They set up a watchman and caught the thieves. They took the thieves to their village and a battle erupted. The thieves were tormented by the people and one thief was scratched and had stripes on his face and back, and his eyes and tail were ringed. This is the corn thief, the raccoon. The people split the upper lip of the Squash thief, the hare-lipped rabbit.[23] This story provides instructions to properly tend the plants, and explains that raccoons and rabbits will steal part of the crops.

The Tuscarora stories differ somewhat, yet the same basic principles of respect for the corn, sharing corn, and conducting ceremonies are evident. "The Origin of White Corn" tells of a people who lived at the foot of high steep cliff.

One day they heard a woman singing from the top of the cliff. The people asked an old man to climb to the top of the cliff and find the meaning of the song. With great difficulty he climbed the cliff and found a young woman. She told him to return in five days to the spot where she had lain. She embraced him, and he swooned. When he awoke she had vanished. He returned in five days and found the corn plant with three ears. He cared for it until harvest time. The next spring he divided the corn equally among the village families and taught them how to plant and care for the corn. He told them corn would become the staple food of their diet.

Another Tuscarora story tells "The Legend of Onenha (the corn)." An old man was very ill and heard voices of women in his dreams. He was instructed to set out a bark bowl to catch the rainwater, which would heal him. He followed instructions and was healed. The women (corn, beans, and squash) taught him the corn songs and dances and told him to return to his people and teach them. They explained that they helped him because whenever he travelled and he found corn, bean, or squash seeds ("we are of course sisters") he would pick them up. They told him that they give thanks when humans plant them and till the ground and they "rejoice" when they hill up the earth around them. They give thanks by "singing and dancing Corn dances." The women instructed him to have human beings offer their thanksgiving through the ceremonial songs and dances. He became ill again and: "Moreover, he saw them—a great company of Corn people, Bean people, and Squash people, and so far as he could see they were in all respects perfect human beings. They danced in a slow, gentle manner in imitation of the waving of the corn stalks as they stand in the ground. Women and girls took part in the dance." The people accepted his words and agreed to conduct the cycle of ceremonies from the planting of corn to the harvest.[24]

Jesse Cornplanter's sketch, "The Spirit of the Corn speaking to Handsome Lake, the Seneca prophet," shows Handsome Lake in a corn field, with the long slender corn leaves of the Spirit of the Corn draping over his shoulders.

> It was a bright day when I went into the planted field and alone I wandered in the planted field and it was the time of the second hoeing. Suddenly a damsel appeared and threw her arms about my neck and as she clasped me she spoke saying, "When you leave this world for the new world above it is our wish to follow you." I looked for the damsel but saw only the long leaves of corn twining round my shoulders. And then I understood that it was the spirit of the corn who had spoken, she the sustainer of life.[25]

In the Code of Handsome Lake the people are instructed to plant the Three Sisters and continue the ceremonies.

The Haudenosaunee, like all cultures, have stories for entertainment, adventure, and amusement. Parker identifies stories for amusement as *gaga'* stating these stories are for fun (and adventure), "rather than religious explanations."[26] "The Boy and The Corn" provides a great adventure story.

An old man and his nephew lived together. The boy never saw his uncle eat so one night he decided to spy on him. The uncle went to a hole and took out a kettle and some corn. He "took a magic wand and tapped the kettle till it grew big." He ate some corn and tapped the kettle until it was again small. The next morning the boy got the kettle out and tapped it. It grew, and grew until it filled the lodge. The furious uncle came back and told the boy the corn was all gone now and it only grows in a dangerous place. The boy, being quite brave, announced he would overcome the monsters and dangers and obtain corn for them. The land of the corn was controlled by women witches and the boy had to outsmart them to get the corn. He escaped in a canoe and returned to his uncle.[27]

"Corn Grinder and the Grandson" tells the story of a young boy raised by his grandparents. He was told never to wander beyond the sound of the corn grinder. He could travel north, east, and west, but never south. Of course, being an adventuresome boy he traveled south to the land of giants. He had great adventures and discovered he was the child of the giants.[28]

Another story relates the adventures of Dagwanoenyent and Gaasyendiet'ha (Cyclone and Meteor) as they interact with a man who eventually slays the monster Dagwanoenyent. In the process of slaying the monster, the man became a human-eating monster, stole a wife, and escaped a monster bear. The story illustrates strength and endurance because he was pursued by the monster bear for ten days and nights. But, with the help of his friend, the Meteor, he outran the monster and arrived at a village. An old man in the village was delighted to do battle with the monster bear and slew him. The old man cooked the bear saying he had been hungry for this food for a long time. The old man "put corn into the pounder and with only a few strokes it became corn meal; then having made bread, he began to eat." The old man took the younger man to see his field of corn. "There were great tall cornstalks with ears of corn on them as long as the man was tall and kernels as large as a man's head. The field extended farther than the eye could see." He showed the young man another field in which all varieties of corn grew and a third field where squash grew. The younger man travelled to another village and was reunited with his wife.[29] This story provides high adventure of cyclones, monsters, transformations, and reunion. In the end, he finds peace, which is shown to him through the corn and squash fields and the reunion with his wife.

We have traced the theme of corn from the Sky World, to the Thanksgiving Address, to the ceremonial cycle, and through oral traditions, which provides us with a background on Haudenosaunee myth, world view, and ceremonies. How does this world view become enacted in structuring the community's daily life? This section will illustrate the concrete interconnection and interrelationship of corn in: 1) Haudenosaunee villages, 2) communal land use/communal work, 3) structuring village life, women's and men's roles, 4) agricultural knowledge, 5) indigenous technology, and 6) diet/nutritional value.

The Haudenosaunee lived in numerous villages, both large and small. Greenhalgh's journey through Haudenosaunee lands from May to July 1677 provides data on the size of villages. He explains that most villages were built on a hill and the cornfields were situated at the bottom of the hill near a river. Greenhalgh cites the abundance of corn at each of the villages and notes that the Oneida village was newly constructed that year and although the fields were planted, the Oneida were buying corn from the Onondaga.[30] Greenhalgh visited a total of fifteen villages in 1677 and it hardly seems likely he could have come in contact with every Haudenosaunee village. Clinton and Sullivan destroyed more than forty villages just among the Cayuga and Seneca in 1779. Either the population increased or the larger villages divided into smaller villages.

VILLAGES	HOUSES
Mohawk—four large, one small village	
Cahaniaga	24
Canagora	16
Canajorha	16
Tionondogue	30
small village	10
Oneida—one large village	100
Onondaga—one large, one small village	140
small village	24
Cayuga	
three villages 100 houses each	300
Seneca—four villages	
Canagorah	150
Tiotohatton	120
Canoenada	30
Keint-he	24

Figure 6.2.   Haudenosaunee Villages in 1677 (Greenhalgh's listing)

The Haudenosaunee people lived in bark longhouses between 50 to 130 feet in length and 16 feet wide. A family area was marked off every ten to twelve feet, with a fire in the center and a family on each side of the fire. Morgan stated that a 120 foot longhouse with ten fires would house twenty families. If he had provided information on average family size it would be possible to estimate population, although he says the largest village contained 3,000 people.[31]

There is a tradition among the Senecas, that at the period of their highest prosperity and numbers, they took a census of their nation, by placing a kernel of white flint corn for each Seneca, in a corn husk basket which, from the description of its size, would hold ten or twelve quarts. Taking the smallest size, and making the estimate accordingly, it will give us the number of Senecas alone at 17,760.[32]

Another population estimate is cited in Dennoville's Journal in reference to Ga-o-sa-eh-ga-aah. The largest town of the Seneca, possibly Ganondagen, the historic site of this village is located near Victor, New York. An old man from Cattaraugus estimated the size of the village as "a kernel of corn for each person made a quantity sufficient to fill five vessels containing one gallon each." In this village the estimated corn destroyed by Denonville in 1687 was 1.2 million bushels, which provides evidence of a well-developed agricultural society.[33]

Living a communal lifestyle required sophisticated social organization. Over the top of the door of each longhouse, the clan animal or bird was displayed. The Turtle, Bear, and Wolf clans were predominant among the Mohawk and Oneida. Onondaga clans included Turtle, Bear, Wolf, Beaver, Snipe, Hawk, Wolf, Deer, and Eel. The Cayuga clans are Deer, Turtle, Wolf, and Snipe. The Seneca clans are Turtle, Bear, Wolf, Snipe, Hawk, Heron, Beaver, and Deer. The clans are determined through the female, thus children inherit the clan of their mother. Each longhouse was a large extended family. A clanmother, usually the oldest woman of the clan, was responsible for giving names to the newborn, selecting chiefs, selecting marriage partners, organizing the agricultural work, and the overall business of the clan.

The Haudenosaunee were famous for their hospitality. When guests arrived they were immediately given food, shelter, and, if necessary, dry clothing. This was considered good manners. Food offered in a Haudenosaunee household was never to be refused as refusal constituted poor manners. In ancient times, one meal per day was served around midday, but a huge kettle of corn soup was always available for people to eat whenever they were hungry. Toward evening hominy was pounded into small pieces and boiled. This dish was available for a snack during the evening or the next morning or to feed guests.

In Haudenosaunee communities there was no starvation, no one was homeless, nor did they ever need to worry about such a possibility because everyone was taken care of as a community responsibility. If there was hunger it was on a village-wide scale due to conditions such as drought, or a poor growing season.

The belief system supporting this generous hospitality was based on the world view, which acknowledged with thanksgiving all the gifts the Creator gave to the people. These gifts, whether from hunting, fishing, gathering, or agriculture ultimately were given by the Creator for everyone to share. These were not gifts to individuals, but to the community, the entire village.

> The housewife announced that a meal was ready by exclaiming Hau! Sedek'oni, and the guest when he had finished the meal always exclaimed with emphasis "Niawe'" meaning, *thanks are given*. This was supposed to be addressed to the Creator. As a response the host or hostess, the housewife or some member of the family would say "Niu'" meaning *it is well*.[34]

It was considered lack of manners, or lack of thankfulness and appreciation of the Creator's gift if one neglected to say "Niawe.'"

The Haudenosaunee system of land use was based on the philosophy that the land was given to the people for everyone to use and work together on the land. The cornfields were communal, but it was also allowable for individuals to plant their own individual fields, but they had to continue to do their share of work in the communal fields. A portion of the communal crops were reserved for ceremonial usage. Planting the corn fields was a community event. Parker describes how the women chose a woman leader who arranged the helpers and supervised the planting. An ordinary planting day consisted of planting the field, with the woman leader, the owner of the field, providing a feast of corn soup. All of this work was "accompanied by singing, laughing, joking and inoffensive repartee, and the utmost humor prevails, topped off by a splash in the water to remove dust and fatigue."

Parker refers to this communal work as a "mutual aid society known as *(In the) Good Rule they assist one another.*" This mutual aid society system extended to many types of work in the village including planting, hoeing, husking, braiding corn, constructing longhouses, making a canoe, harvest, hunting, assisting when someone was ill, and ceremonial events. The merriment of the husking bee was described by Parker as a joyous time when everyone helped to husk corn after the harvest. The men would bring their water drums and horn rattles to provide songs for social dancing. The older people often told stories to the children.

> The "bees" were often conducted out of doors under the white moonlight. A roaring fire of sumac brush or logs tempered the crisp air of the night and left it sufficiently invigorating to keep up spirit and keep the workers active. There was nothing

unhealthful in these night carnivals where the smell of the corn plant, the breath of the pines blown by the autumn wind, the smoke of the fragrant burning wood and the pure merriment of the workers and the knowledge of good work furnished the sole exhilaration.

Parker shares the good feelings of the community working together. These "work bees" are healthy experiences for community cohesion as Parker witnessed in the early 1900s.[35]

Peter John, an Onondaga, remembered going to work bees around 1866–71. John said work bees were frequent and he took his own hoe and eating utensils for the corn soup or corn bread given for his share of the work. "When one person's corn patch was finished they would go on to the next."[36] The Haudenosaunee recognized the unity of working cooperatively and enjoying the work and each other. The people established rules for their social organization that were based on thankfulness and sharing.

The men were responsible for the slash-and-burn method of clearing the fields. The trees were girdled in the spring and the following year the dead trees and underbrush were burned. Parker mentions that in the midwinter thanksgiving it is asked that "dead branches may not fall upon the children in the fields."[37] Men were responsible for hunting, fishing, and warfare along with ceremonial duties.

The fields and the crops belonged to the women who were in charge of the staple crops of corn, beans, and squash. The women organized the planting and the first and second hoeing or hilling up of the corn. The women gathered the wild plants for food and medicine. One prohibition was that a women in her menstrual cycle should not participate in the planting.

Lewandowski provides an analysis of agriculture and the Haudenosaunee:

> The "whole world" of the Haudenosaunee "revolved as one with the agricultural cycle" as can be seen in Mary Jemison's story. At one point she mentions that she bore a child "at the time that the kernels of corn first appeared on the cob." Her child dies, and Mary was taken so gravely ill, that many thought she would die, but her "complaint took a favorable turn, and by the time that the corn was ripe (she) was able to get about." For the Seneca, crops were the clocks that traced their course through the year. . . . The Three Sisters cultural complex consisted not only of agricultural and nutritional

strategies but of a body of stories, lore, ceremonies, and customs that touched every facet of their lives. The Three Sisters complex was not "progressive" or expansive in most senses; it was conservative and enduring. It seems to have returned good, steady yields to "leisurely" labor and to have been sufficient to *support a culture that used land without spoiling it.*[38] (emphasis added)

The male and female roles were complementary in Haudenosaunee lifestyle. There was a balance inherent in these roles which worked to keep the village in harmony.

The Three Sisters formed a "symbiotic complex, without an equal elsewhere." Haudenosaunee cultural beliefs about the Three Sisters can be confirmed as a sound agricultural system. The cornstalks serve as a support for the beans to climb, the roots of the beans "support colonies of nitrogen-fixing bacteria,"[39] and the ample squash leaves keep moisture in the hills and provide ground cover to keep down the weeds. The agricultural system of the Haudenosaunee was based on scientific knowledge including: knowing when to plant, using specific plants as a "medicine" to soak the seeds before planting, using fertilizer, hill planting, and intercropping along with ceremonial practices. Planting usually took place when leaves on the oak tree were as big as a red squirrel's foot, or when the Juneberry blossomed.[40]

The seed corn was soaked in a mixture of specific plants and water before planting. Soaking seeds in these herbs created a "poison for crows and other field pests. . . . A bird eating this "doctored" corn becomes dizzy and flutters about the field in a way which frightens the others."[41] In 1916, Waugh found that knowledge of these "medicines" continued to exist among Cayuga, Onondaga, Tonawanda, and Alleghany peoples. The roots of the *"Phragmites communis,* a tall, reed-like grass growing in marshes; and *Hystrix patual,* or bottle-brush grass, also growing in the low land" were boiled in water and when cooled, the seed corn was added to soak for an hour. The water was drained and the seed corn placed in a basket to begin sprouting. It was said the root was taken to the longhouse and everyone who intended to plant could take some of the root to prepare their seed corn.[42] In the Code of Handsome Lake, reference is made to a medicine, "o'sagan'da' and o'sdis'dani," in which the seeds were to be soaked before planting.[43]

Corn was planted in permanent hills that were evenly spaced about three feet apart in distinct rows also three feet apart. Hills provided a concentrated area which the people kept fertilized with fish for optimum plant growth. Hill planting prevented soil erosion which is a lesson modern day agriculturalists are beginning to recognize as a way to preserve valuable topsoil. The fields were generally on the "rich bottom land soils which we know to be most fertile."[44]

The Haudenosaunee practiced intercropping corn with beans and squash. Among the Seneca, squash and beans were planted in every "seventh hill because it was thought that the spirits of these three plants were inseparable."[45] Another method was to plant the beans in the hills two to three weeks after the corn was planted. Squash or pumpkins were planted between the rows. Lewandowski cites accounts of four to six corn seeds per hill and a "ring-planting pattern of beans and squash." The first hoeing began when the corn was a "span high." and the second hoeing, called the hilling up, took place when the corn was knee high.[46]

The planting ceremony lasted one day and had two purposes, first to thank the Creator for the return of the planting season, and second, to ask for a good harvest. Tobacco and wampum were offered "to the spirits of growth and to the pygmies, Djo ga'o." In the Thunder Ceremony the Creator was "asked to remember the fields with a proper amount of rain and prevent the maize fields from parching."[47] The Grand River Planting Ceremony, described by Chief Gibson, says that when the people assembled in the Longhouse, the speaker made a

> speech to the effect that a good number of people still have the privilege to plant again. He gives thanks to the corn, makes an offering of...native tobacco, and continues at some length to thank all green things, or whatever grows on earth in spring.

For the dish bowl game the people wager seeds of corn, beans, squash, pumpkin. The Towii'sas women's society with their small turtle rattle sing their songs to thank the Mother Earth and the Three Sisters. The men sing alternately with the women clapping their hands. The women then stand and the leader of the woman carries the braid of corn, and seeds, and sings while the other women form a line behind her as they go around the Longhouse three times. The winners of the dish bowl game are given the seeds.[48]

The cultural, agricultural, and social dimensions of Haudenosaunee society were intricately interconnected and intertwined in a complex whole that made infinite sense for them.

In selecting the fields for land management and soil fertility, the Haudenosaunee selected fields near rivers, which were naturally fertile. They rotated field usage rather than crops. When a new field was prepared by girdling the trees in early spring and burning the dead trees that fall or the following spring, the ashes provided nutrients that enriched the soil. Returning the cornstalks to the hills provided organic material, which improves the soil in the concentrated area of the hill and prevents soil erosion.[49] When an area of soil was depleted, the entire village would be moved to another location and the land left to fallow, that is, to restore itself with natural vegetation.

In addition to the Three Sisters as a cultural belief, interplanting promotes sound agricultural principles. The beans "are able to fix nitrogen, meaning that they take atmospheric nitrogen and turn it into a form that's available to plants."[50] The corn provides support for the beans, and the squash provides weed control through ground cover.

Lewandowski provides a scientific explanation of the root systems. The corn has a shallow root system, beans an intermediate root system, and squash/pumpkins a deep root system. "Rhizobial symbionts [beans] fix atmospheric nitrogen and make it available to squash and corn, which lack this ability." This hill system suppressed weeds, and limited evaporation of water from the ground's surface.

> The Three Sisters weave a lush carpet of vegetation of variously shaped and mounted leaves capable of capturing the solar energy falling on the patch, of reducing rain drop impact and thus soil erosion, and of utilizing water and nutrients with greater efficiency. All three plants have passed through a long process in which complementary traits were chosen for their most beneficial association. . . . The nutritional strategy embodied in the Three Sisters was worked out over at least seven thousand years.[51]

Mt. Pleasant discusses the types of corn and varieties of corn developed through scientific experimentation: "they had a clear understanding of corn breeding. They understood . . . that if they wanted to keep pure strains of their varieties with separate characteristics, they need to keep isolated plantings of those different corn varieties."[52] Selection of seed for planting was a crucial form of knowledge which had developed over thousands of years of scientific observation and experimentation. Parker provides a photograph and lists the varieties of corn, including soft, flint, sweet, and pod corn or sacred original corn.[53]

According to Lewandowski, the Seneca knew how to sprout seeds indoors in bark containers and transplant seedlings in the fields. Solar energy was utilized by placing large stones near the young plants to collect heat.[54]

The Haudenosaunee were using sound agricultural principles that had been developed over thousands of years into a complex agricultural system. Mt. Pleasant cites several instances in which researchers at Cornell are just now discovering the Haudenosaunee agricultural practices of no-till, ridge tillage, and intercropping were based on a sustainable system of agriculture. "They used a system of corn cultivation that was based on sound agronomic principles, and those are the same principles that we still use now in 1988 for growing good corn."[55]

Indigenous knowledge was utilized to produce tools from natural materials to meet specific purposes. Several of these technologies, developed in ancient times, continue to be utilized in contemporary times and cannot be replaced with current technology.[56]

The digging stick resembled a short hoe and was made of a large flat bone or piece of wood attached to a wooden handle. This tool was used to break the ground for the planting hills and to hill up the soil around the corn during the second hoeing. Planting baskets to carry the seed to the field were made from elm bark or of splints. These baskets were carried or could be tied around the waist with a belt, thus leaving both hands free for planting. The husking pins were made of bone or antler and looked like awls. A leather loop went around the middle finger to hold the husking pin, which was used to split and remove the husk from the corn.

Utensils for preparing foods and eating consisted of bark or wooden bowls, and elaborately carved spoons with a bird or animal on the handle of the spoon. A larger bark bowl was made from bark peeled from the tree in the spring and bent into shape and bound around the edge with a "hoop of ash sewed on with a cord of inner elm or basswood bark." These bowls were used in preparing corn bread and measure one to two feet in diameter and four to nine inches in depth.

Everyday eating bowls were smaller and made of maple, oak, or pepperidge knots. They were carved and polished with a dye "solution made from hemlock roots," which with scouring gave a "high polish," and grease gave it "an attractive luster." These bowls could be plain or could have carvings such as a bird's head. Bark bowls were so sturdy that they could be used for cooking by putting heated stones into them to boil foods. Wooden bowls were made of pine or maple.

Elaborately decorated and carved wooden bowls have been found that were used for ceremonies and feasts. Parker describes a feast bowl that has a handle carved into a beaver tail. The beaver tail is a symbol of peace that dates from the Great Law of Peace. Another bowl is divided into five sections by yellow lines, with beaver tails and the names of the Mohawk, Oneida, Onondaga, Cayuga, and Seneca painted in them. This bowl symbolizes the departure of some of the Haudenosaunee to Canada after the American Revolution.

Baskets were made from splints from the black ash tree. This involved an elaborate process of locating the correct tree, cutting it down, pounding the tree for splints, cutting the splints to the various widths for different baskets, and weaving the baskets. Baskets served many useful functions in preparing corn. Each weaving design suited the function of the basket. The sifters for hominy, meal or flour, and ash were a tight weave which allowed only the fine grains to fall through. Carrying baskets were a necessity when harvesting the crops. A looser weave was used for the hulling or corn washing baskets. This permitted the corn, after cooking with wood ashes, to be dipped in a stream to wash away the ashes and hulls. The hulling basket and carrying baskets continue to be made at the present time.

The wooden mortar and pestle were essential tools. When Sky Woman was falling from the Sky World, the Fire Beast gave her corn and the mortar and pestle. The mortar is made of the pepperidge or black oak tree. A section of the trunk about twenty inches in diameter and twenty-two inches long was selected. The hollow in the top of the mortar was made by burning the top surface and then scraping it out. This process was repeated until the bowl-like hole was about a foot deep. The pestle was about forty-eight inches long and carved out of hard maple. The center area is smaller where the person can grasp it with both hands. The corn pounder was used to pound corn into smaller pieces for boiling and into fine grain flour for breads.

Wooden paddles, both wide and narrow, were carved of wood to use in stirring dishes and to place the corn bread loaves into the boiling water and to remove them when done. These paddles could be very plain or could be carved with fancy designs depicting birds or animals.

In ancient times, clay pottery was made and decorated with distinctive Iroquois designs. The round bottom pottery was placed on four stones over the fire, or was suspended on poles to boil foods.

Corn husks were tightly woven for salt bottles, used to make corn husk dolls, woven into mats and footwear, and woven into ceremonial masks.

Each of these items of indigenous technology used nature's gifts for the raw materials, and human ingenuity, human intelligence, provided the knowledge to make functional items necessary for humans to live comfortably.

Corn was braided into long strings and placed in the rafters of the longhouse to dry. The smoke from the cooking fire dried the corn and kept pests away. "It hung like a tapestry the whole length of the cabin."[57] Reports indicate that enough corn was stored to last three to four years, with extra corn used for trading. If any of the Haudenosaunee nations had an emergency they could count on the other Haudenosaunee nations to assist them with corn.

After the corn was shelled it was placed in bark barrels which were stored in an entryway or additional section of the longhouse for that purpose. Elm bark containers of parched corn, shelled beans, and sun or fire-dried squash were stored under the bed. One reference was made to round bark granaries which were placed on "elevations, piercing the bark from all sides so that the air" would get in to dry the corn and prevent spoilage. Chief Gibson described another type of storage which was constructed from small posts no more than six inches in diameter. These poles were planted one and a half feet deep in the earth in a circular hole. The hole was then filled with earth and packed down. The corn, on the cob, was placed in the structure and elm bark pieces served as a cover. Another pole was placed over the top and "tied down with strips of basswood inner bark."[58]

Underground pits were dug four to five feet deep and lined with bark or grass or hemlock branches. Corn was put in the pit and a top was made of bark and the whole pit was then covered with soil. In the early 1900s these underground pits were found throughout Haudenosaunee lands. Corncribs were common throughout Haudenosaunee lands. And, according to Parker have been "little improved upon by white men." Several photographs by Parker and Waugh document both the pole construction and the later use of slats and boards.[59] Inside the corn crib poles were placed across the top beams and the braided corn was hung on these poles.

The Haudenosaunee daily diet consisted of foods obtained through hunting, fishing, gathering, and agriculture. The typical Huron diet included 65% corn, 15% beans, squash, and pumpkins, 10–15% fish, and 5% meat. This diet required 1.3 pounds of corn per person per day. The Huron and Haudenosaunee diets were quite similar although a percentage breakdown has not been found for the Haudenosaunee. The above percentages do not include the many varieties of foods gathered from nature including many varieties of berries, maple syrup, fruits, nuts, mushrooms, and greens. Analysis of the nutritional components of the Three Sisters reveals:

> Corn is an excellent source of carbohydrates (74%), protein, (9.2%), and polyunsaturated oils (2%). . . . Beans . . . provide ample lysine and tryptophan to bring corn's protein into balance. Dry beans add more protein (22%) than fresh snap beans (10%). Judging by recipes, the Seneca used many more beans in a dry form. Squash . . . is a good source of carbohydrates and sugars. If, however, squash is selected for seeds, more protein is added to the diet.[60]

The corn and beans together provide a complete protein, and squash added carbohydrates and sugars.

Typical foods prepared from corn include the staples of corn soup and corn bread, both prepared with beans. Other dishes include parched corn mush, hominy, boiled corn, wedding bread, roasted corn, and many variations of these dishes, which included the addition of nuts, berries, and squash.[61]

The preparation of corn for both corn soup and corn bread begins with the same process to remove the hard outer hull of the corn, which is called washing the corn. White corn is added to boiling water and to this is added sifted wood stove ashes, which turns the corn a bright orange. The ashes form a lye solution that softens the outer hull. When the corn turns white again and the hulls soften, usually after thirty to forty-five minutes, the corn is taken to a nearby stream or a large tub of water to be washed. The corn is poured into the corn hulling basket and dipped into the stream to wash away all the ashes and hulls. When the corn is thoroughly washed, it is added to a fresh pot of boiling water and boiled for several hours.

The preparation varies somewhat from cook to cook. Cornplanter gives exact measurements of 2½ quarts of white corn to 1½ quarts of ashes and boiling for 1½ hours with repeated washing.[62] Another method is to boil the corn until it begins to swell then add one quart of sifted hardwood ashes per gallon of water. It is boiled until "the black pips of the corn loosen and may be seen floating about in the kettle, and the hull slips easily when the grain is rolled between the fingers." The corn is then washed and boiled one more time and rinsed before the final cooking.[63]

Washing the corn with ashes softens the hard outer layer and this process "enhances the nutritional quality" of the corn, which provides an "overall better balanced food."

> Thus, as long as maize is the major component of the diet, then cooking techniques in which alkali and heat are used clearly enhance the balance of essential amino acids and free the otherwise almost unavailable niacin.[64]

When corn soup is prepared for the Green Corn ceremony it is washed in the same manner but instead of boiling it again, it is placed in the corn pounder with a little water and pounded. It is then put in boiling water with berries and a little sugar (maple). Other variations of this include adding beechnuts, apples, or sunflower seeds.[65]

Making corn bread begins with the same process of washing the corn with wood ashes. The corn is dried and pounded into corn flour. Hot water and beans are added to a portion of the flour, which is then formed into a circular cake or wheel about six inches in diameter and about two inches thick. The hot water coagulate(s) the starch, which makes the flour hold together as the circular loaf is formed. The corn bread maker puts her/his hands in cold water and rubs them over the corn bread to give the loaf a glossy finish. The cake is placed on a wide corn paddle which is used to stand the circular cake on its edge in a kettle of boiling water. The corn bread boils about an hour and is done when it floats to the surface. Many varieties of corn bread are made by substituting blueberries, sunflower seeds, nuts, or any type of berry, for the beans. Strawberries are always added for wedding bread. In ancient times this wedding bread was not made into a cake or wheel but into two smaller loaves which were tied together with corn husks and boiled.

Parched corn mush is another favorite dish. The white corn is parched in a heavy pan and ground into flour. The flour is sifted and then cooked with boiling water. This cereal type of dish is eaten with maple syrup and sometimes salt pork. In ancient days this parched corn was carried by hunters on long trips and mixed with maple syrup and water.

In late August, the ears of corn are husked and roasted over an outdoor fire. The roasted kernels are scraped from the cob and dried. During the winter, roast corn is boiled with water and salt pork for soup.[66]

Another food, ogon'sa' or baked green corn, is prepared while the corn is still in the milky stage, that is, when the corn is scraped off the cob and is full of moisture. After scraping, the corn is either sun-dried on trays or dried over the fire; in contemporary times it is dried in the oven. When used later in the year it is boiled. Parker's description of ogon'sa' is rich with details of the time period (1910):

When the milk has set, Tuscarora and sweet corn is scraped from the cob and beaten to a paste in a mortar. This should be done just before the evening meal. After the housework is finished, the housewife lines a large kettle with basswood leaves three deep. The corn paste is then dumped in up to two thirds the depth of the vessel. The top is smoothed down and covered by three layers of leaves. Cold ashes to a finger's depth are now thrown over the leaves and smoothed down. A small fire is built under the kettle which hangs suspended from a crane or tripod. Glowing charcoal is placed on the ashes at the top. The small fire is kept brisk and the coals at the top renewed three times. The cook may now retire for the night if her kettle hangs in a shielded place or in a fire pit. In the morning the ashes and top leaves are carefully removed and the baked corn dumped out. The odor of this steaming ogon'sa' is most appetizing and it is eaten greedily with grease or butter. For winter's use the caked mass is sliced and dried in the sun all day, taken in at night to prevent dew from spoiling and dogs or night prowlers from taking too much of it, and set out again in the morning to allow the sun to complete the drying. The ogon'sa' is then ready to be stored away for the winter. When ready for use the winter's store of ogon'sa' was taken from storage and a sufficient quantity for a meal thrown in cold water and immediately put on the stove. Boiling for a little more than a half hour produces a delicious dish. Ogon'sa was one of the favorite foods of the Iroquois and remains so to this day. An Onondaga or Seneca can hardly mention the name without showing that it brings memories of the pleasant repasts that it has afforded.

In recent years the corn paste is prepared with a potato masher in a chopping bowl, or by running the corn as cut from the cob through a food chopper. Baking is done in shallow dripping pans in the oven. The food so prepared, however, lacks a deliciousness that makes the older method still popular.[67]

Many corn dishes were made by pounding the corn in the corn pounder and sifting the corn flour through the finely woven sifting baskets. This method continues to be utilized for ceremonial foods in contemporary times. Foods from corn were also boiled or wrapped in corn husks and baked in ashes. Parker provides the Seneca name and detailed descriptions on the preparation of the following foods: leaf bread tamales which were wrapped in corn husks and boiled; boiled green corn or corn on the cob; fried green corn; succotash made from scraped green corn boiled with beans; baked cob-corn in the husk; baked scraped corn; cracked undried corn; baked corn bread, which would not spoil and was taken along when travelling; corn soup liquor, which is a beverage made from the water in which the corn bread was boiled; early bread made from green corn; early corn pudding; dumplings; hominy, which is pounded and sifted or winnowed and boiled; dried corn soup; nut and corn pottage, a meal made of pounded nuts and parched corn meal; corn and pumpkin pudding; samp, which is like corn bread soup with berries or meat; corn pudding or parched corn mush; roasted corn hominy; parched corn coffee; and pop corn pudding, which was corn popped in a clay kettle and pounded then mixed with oil or maple syrup.[68]

The Haudenosaunee were a healthy people who feasted on a nutritionally sound diet of corn, beans, squash, gathered fruits, nuts, berries, mushrooms, greens, fish, venison or bear, or meat from small game animals.

Haudenosaunee Culturally Specific Paradigm

"All Seneca dances are counter-clockwise."[69] The Haudenosaunee observe nature's direction in their dances. Pole beans wind around the pole in a counterclockwise direction. "Nature moves counter-clock wise in the northern hemisphere and you will find dances are counter-clockwise in agricultural societies in this part of the hemi-sphere. The Lakota dance in a clockwise direction because their world view is based on the animal world and the cycle of the sun."[70]

The following circular chart illustrates the components of Haudenosaunee world view around the thematic focus of corn and the interrelationships of these components in daily life. The chart is to read in a counterclockwise direction.

Corn first appears in the Sky World villages. The Thanksgiv-ing Address, which includes the Three Sisters, is acknowledged in the epic narratives, ceremonial cycle, and oral traditions. There are

WORLDVIEW   **ENACTMENT IN DAILY LIFE**

Figure 6.3. Haudenosaunee Culturally Specific Paradigm

many stories that explain the importance of corn in Haudenosaunee culture. Village life centers around the agricultural economic base of corn. Indigenous technology was developed to furnish tools and utensils necessary for agricultural life. Slash and burn clearing of fields, hill planting, and intercropping formed a sound agricultural system. Corn, washed with ashes, and beans, and squash formed a nutritionally balanced diet. Fishing, gathering, and hunting supplemented the corn staple by providing a variety of flavors and nutrients to the diet.

The ceremonies brought the people of the village or community together to express their gratitude and thankfulness for the Creator's gifts. Thus we see the world view, (left side of the chart), enacted in daily life, (right side of the chart) achieving harmony, i.e. balance.

The Haudenosaunee world view is extremely ancient and each element is interdependent within a cultural complex.

The Three Sisters is where agriculture and horticulture, and human culture meet. Is it gardening on a large scale or farming with a human face? In any case, it is a cultural complex, with distinct but inter-related factors, in which the functions of planting, harvesting and eating are more than simple biological necessities; they are elements of a well-recognized sacrament. The Creator who gave the Three Sisters, and the guardian deities, the Sisters themselves, were constantly brought to mind through a series of seasonal festivals.[71]

With this background on the healthy lifestyle of the Haudenosaunee, which combined world view with oral traditions, ceremonial practices, and sound agricultural principles, we can look at this culture in a holistic manner.

Chapter 7

The Interaction of Corn and Cultures

The interconnectedness of Haudenosaunee world view and way of life discussed in the previous chapters is part of a much larger pattern of corn and culture throughout the western hemisphere.

The first section of this chapter presents two perspectives on the origins of corn and expands the thematic focus on corn from a specific culture to a view of corn in many indigenous cultures throughout North, Central, and South America. Geographic areas of the Americas are examined to provide a view of corn including cultural beliefs, indigenous agricultural practices, and the importance of understanding the nutritional value of corn. The last portion of this section looks at how Columbus dispersed corn to Europe, Asia, and Africa where it was incorporated into these cultures in several ways.

The second section examines the significant role of corn in the interaction between cultures within the northeastern United States beginning with Squanto, the fur trade, and analyzing corn's impact on the early settlers. This two-way cultural interaction, in which both cultures obtained positive and negative factors from each other, occurred during the exchange of agricultural knowledge, trade systems, wars, and government policies, as corn became a major crop of the United States.

The third section examines the role of corn in wars against the Iroquois by looking at the vast quantity of corn found by the French during Denonville's 1687 attack on Seneca villages and Clinton and Sullivan's "scorched earth campaign" in 1779. The destruction of Haudenosaunee corn was a military practice to destroy or weaken American Indians throughout colonial history.

The fourth section provides a view of United States government assimilation policy, which supported the Quakers' attempt to make "white farmers" out of "Indian farmers" in the late 1700s and early 1800s. The loss of land after the American Revolution and during the treaty period was devastating to Haudenosaunee economy.

The fifth section surveys the time period after 1800 into the early 1900s when Americans were eating corn, planting corn, and building the U.S. economy with corn as a major crop.

Origins of Corn in the Western Hemisphere and Dispersal to Europe, Asia, and Africa

American Indians, prior to contact, were primarily agriculturalists with hunting, gathering, and fishing supplementing their diet. Corn was a staple crop from what is now known as southern Canada, throughout most of the United States, Central America, and into South America. This section provides an overview of the origins of corn, reports of corn by early explorers, and culturally specific stories that show the interdependent relationship of corn and human beings. A summary shows consistent agricultural practices and looks at the nutritional value of processing corn with alkali. The last part of this section presents the dispersal of corn to Europe, Asia, and Africa, examining the different ways the gift of corn was incorporated in these areas. Thus, corn is examined on a worldwide level.

Origins of Corn

There are two perspectives with which to examine the evidence of the origins of corn throughout the Americas, archeological theories and the indigenous or *in situ* perspective. The indigenous perspective supports independent, or *in situ,* origins of corn. Throughout the Americas indigenous peoples have cosmological beliefs about their origins and how corn came to their people. These oral traditions, which have been handed down from generation to generation, contain specific knowledge about the origins of corn. The corn itself differs from the northern climate of New York State, to the near-desert conditions of the southwest, to the highlands of Mexico because the corn seed fits the climatic conditions of its environment. Therefore, *in situ* origins, that is, specific origins within specific land-based cultures, independent of each other, form the basis of beliefs surrounding the origins of corn.

The other perspective, based on archaeology, maintains that corn originated as the wild grass teosinte in the Tehuacan valley of MesoAmerica around 4500 B.C. and was dispersed north and south.

> Archaeologists have recovered prehistoric corn from five caves—Coxcatlan, Purron, San Marcos, Tecorral, and El Riego—in the Tehuacan Valley, which show the evolutionary development of the plant over a period of sixty-five hundred years.[1]

Corn existed in MesoAmerica in 7000 B.C. By 4500 B.C. the presence of corn, beans, squash, and other crops provide evidence that people lived in permanent villages utilizing slash and burn methods to clear their fields. Evidence of corn in pre-Colombian times has confirmed that corn was grown by the Anasazi in the southwest, the Moundbuilders of Ohio, the Illinois, as well as indigenous peoples in the southwest, Central America, and South America.

Archaeological evidence does not include the beliefs of the people, thus ignoring or eliminating indigenous beliefs. This study includes multiple perspectives. Hurt provided a healthy view saying that "In all probability, though, no one will ever know when the first cultigens were adopted or by which people." Therefore, the two perspectives exist, the literature and textbooks support the evolutionary theory, but indigenous peoples continue, as they always have, to believe traditional stories of the origins of corn, beans, and squash, as transmitted to them from generation to generation into antiquity. However, archaeological work is important because each new discovery confirms the antiquity and widespread existence of corn, beans, and squash throughout the Americas.

Corn throughout the Western Hemisphere

Corn was the staple crop throughout the western hemisphere. Beginning with the United States, corn was grown in the northeast, southeast, midwest, Great Plains, and the southwest. Early explorers, missionaries, fur traders, trappers, and military personnel documented finding corn throughout the Americas.[2]

In the northeast, early explorers noted seeing corn among the Huron, Fox, Illinois, Algonquians, Susquehannocks, Algonquins,

Delawares, Mahican, Narragansetts, and Abenakis.[3] Cartier saw
corn as he explored the St. Lawrence River area in 1534, and as
Champlain traveled throughout the northeast in the early 1600s
he saw corn fields from Cape Cod to Lake Champlain, New York.[4]
In 1609, Henry Hudson cited enough drying corn to fill three ships
plus corn still in the field.[5] In 1612, Lescarbot described hill planting and intercropping of corn and beans among the Indians of
Maine, Virginia, and Florida.[6]

Harmen Meyndertsz van den Bogaert, a Dutchman, travelled
into the Mohawk and Oneida territory from Fort Orange, now
Albany, New York, in 1634. What he saw documents reports by
many explorers throughout the Americas.[7] Bogaert went to discuss
the price of furs with the Mohawk and Oneida who had recently
traded with the French, who offered a better price for the furs. He
found along the trail a number of little huts built for travellers and
hunters. He met people who were so frightened that they ran away.
Van den Bogaert ate a "small loaf of bread with beans," which he
found in their bags. They came to a village of thirty-six houses
which was built on a hill. He describes longhouses of bark that
were eighty to 100 steps long and twenty-two to twenty-three feet
high. He saw items made from iron, which is evidence of trade.
Most of the people were out hunting bear and deer, and he saw
that the houses were full of corn and some held 300 to 400 skipples
(.764 bushel). Baked and boiled pumpkins were also eaten at this
village. The chief of the village "was living one quarter mile from
the fort in a small cabin, because many Indians here in the castle
had died of smallpox." They cooked a turkey and put the grease
from the turkey on their corn and beans, and he mentions beaver,
salmon, venison, and bear. He provides detailed descriptions of the
many small villages they passed by and those where he stopped
and the wonderful hospitality afforded them at each stop. Awls,
axes, scissors, needles, kettles, and knives were exchanged in these
villages for furs and wampum, and to confirm negotiations. Van
den Bogaert spotted French axes, shirts, coats, and razors, which
indicated they had been trading with the French. He described the
fourth castle (a castle indicates a palisaded village), on the other
side of the river which was full of corn and beans. They were given
supplies for their trip that included corn bread cooked with nuts,
chestnuts, blueberries or sunflower seeds. At one village they saw
houses with sixty or seventy dried salmon and said they could
catch in that river 600–800 salmon in one day. In constructing a
map to show the locations of villages, they used kernels of corn and
stones. The Mohawks and Oneida had a healthy diet of corn, beans,

squash, nuts, berries, salmon, and venison, bear, and beaver. The trade items visible in the villages, and those items Bogaert traded, are evidence of considerable interaction between the Mohawk and Oneida with the French and Dutch.

The Indigenous Peoples of the Northeast were exchanging corn and agricultural techniques with the settlers who were trying to convert them to Christianity.

> As showing the importance of corn to the Indians, we may note that Rev. John Campanius, in his Delaware and Swedish translation (1696) of the Catechism, accommodates the Lord's Prayer to the circumstances of the Indians thus: instead of "give us our daily bread," he has it, "a plentiful supply of venison and corn."[8]

Southeast

In the earliest records of the southeast, De Soto, in 1539, observed that the Apalachee and Timucua Nations of northern Florida were agriculturists. In Virginia, Captain John Smith saw corn, beans, and squash planted in hills. A 1614 treaty between the Chickahominy Indians of Virginia and early settlers included the provision that the Chickahominy would supply the colonists with one thousand bushels of corn annually, in exchange for iron hatchets. The European settlers noticed the consistency used by Indigenous peoples in selecting the best soil and took over these areas. In the lower Mississippi River area, the Natchitoches of Louisiana planted early and late varieties of corn.[9]

The Cherokee, Choctaw, Chickasaw, and Creek Nations raised corn, beans, and squash. The Choctaw, in 1770, continued their agricultural practices of corn, beans, and squash and other crops, while engaging in procuring deerskins for trade. The Choctaw planted in large communal fields, and had smaller family gardens, which belonged to the female lineages.[10]

The Mikasuki, one group of the Seminole of Florida, raise corn and have planting and harvest ceremonies. This story relates the origins of corn among the Mikasuki:

There was a Grandmother, and two grandsons who lived together. The boys hunted for her every day, but decided they were tired of eating only meat so the Grandmother said she would prepare something new. They brought home a deer and she put chunks of venison in the boiling mixture, which was corn. The next day she added wild turkey to corn grits. The next day the boys provided muskrats, which she added to dried cracked corn. The boys were curious about this new food their Grandmother cooked. The younger brother decided to hide and spy on his Grandmother to find out about this new food. She went to a storehouse, spread out a dried deer skin and placed a wooden bowl on it. She rubbed her hands down her sides and corn tumbled from her body onto the deer skin. The younger brother explained what he saw to the older brother who said, "That's impossible! Why would she want to do that? Why does she think we want to eat our own grandmother? Only Caddos and Tonkawas are cannibals! Mikasukis would never eat anybody." They were unable to eat or swallow the corn that evening. The Grandmother fell to the floor, weak, saying since they had found her secret she must leave them. They were dismayed.

She told them to bury her in a field that never floods and cover her with good soil and build a fence of sticks around her and keep people away. She told them that the green plants would grow with tassels and to pick the grain in the fall, dry it, store it in a clean dry place. She instructed them to find wives and plant corn the next spring in hills. They were told to plant four grains of corn and four beans to every hill which would grow together, and, "you may plant their little sister, the squash, between the rows . . . as long as you have the corn, I will be with you."[11] Note the view of the Mikasuki towards the Caddos and Tonkawas, as a subtle form of Indian humor. The presence of the Grandmother and two boys is similar to Haudenosaunee cosmology.

Midwest

> Cahokia farmers grew corn, squash, and beans; hunters set out to bag deer, ducks, geese, and swans; lakes and streams yielded tons of fish.[12]

At this ancient site, the flourishing Moundbuilders constructed Monk's Mound with a sixteen-acre base, which surpasses that of Eygypt's Great Pyramid. The farmers of Cahokia used a flint hoe to cultivate their vast cornfields, which fed 30,000. Stuart speculated building fortified farming communities became necessary, as people depended on agriculture for their livelihood.

In 1699, an explorer stated that the Quapaw tribe on the Illinois River and the Sac Nation along Rock River cultivated corn. Chief Black Hawk said, "the land being good, never failed to produce good crops of corn, beans, pumpkins, and squashes. We always had plenty—our children never cried with hunger."[13] The French explorer Nicholas Perrot, between 1680–1718, remarked that the primary crops of the Indians in the "Old Northwest" were corn, beans, and squash. Perrot claimed corn was planted "continuously from the present location of Fort Wayne, Indiana, to Lake Erie." Sturtevant cites Marquette (1673), Alouez (1676), and Membre (1679) as documenting that corn was grown by the Illinois Indians. La Salle, in 1679, took about forty bushels of corn from the vast quantities of corn of the Illinois Indians.[14] Hennepin (1680) saw corn from "Niagara to the Mississippi river." In 1794, a letter from General Wayne states that the Delawares of Ohio grew corn in immense fields.[15] Throughout the central portion of this country we find agriculture based on corn, beans, and squash.

Great Plains

The Great Plains peoples are not generally thought of as agriculturalists, but rather as a nomadic buffalo hunting society. However, in 1804 Lewis and Clark found the Sioux of the Upper Missouri raising corn, beans, and potatoes.[16] A Lakota (Sioux) story tells of a young couple asking for a blessing:

> Day after day, they watched the plant and it grew and grew and seemed to form arms somewhat like a human, but they didn't understand what it was. When it grew up, it had a yellow tassel; it was the corn plant. It was not the blessing they sought but it was what transformed their culture from one state to another. Corn was the agent of transformation.[17]

This ancient memory of corn among these buffalo hunters illustrates that the Lakota grew corn before the introduction of horses.

An agricultural environment flourished in Nebraska among the Pawnee. This agricultural society changed to a more mobile society that hunted buffalo on a wider scale after the introduction of the horse around 1700. Nevertheless, those Indian nations who remained agriculturalist became an important source of food, a trading center, for hunters.[18]

The Mandan and Hidatsa of North Dakota were the most prolific agriculturalists in the area. Buffalo Bird Woman was born about 1839. Gilbert and Wilson interviewed her in 1906, because, "he wanted to tell about Indian life from the Indian point of view." *Buffalo Bird Woman's Garden* provides the feeling that you are sitting at Buffalo Bird Woman's kitchen table listening to her talk about her life. She gives intricately detailed accounts of the agricultural cycle of the Hidasta. She described a typical planting day which began very early in the morning. By ten A.M. they had planted ten rows or about 225 hills of corn. The Mandan/Hidasta did not plant the corn and beans in the same hill but alternated a row of hills in corn, a row of hills in beans, and another row of hills in corn to make a pattern. Squash was planted in hills to separate or define borders of the garden from the neighbors' garden. Sunflowers served the same purpose. She shares the logic, the experiments, handed down to her about how, when, where, why to plant. She discusses the importance of keeping the varieties of corn separate, because they "travel," or mix if planted together. They built watchtowers to guard over the corn fields. Buffalo Bird Woman described her feelings about corn:

> We cared for our corn in those days as we would care for a child; for we Indian people loved our gardens, just as a mother loves her children; and we thought that our growing corn liked to hear us sing, just as children like to hear their mother sing to them. Also, we did not want the birds to come and steal our corn. Horses, too, might break in and crop the plants, or boys might steal the green ears and go off and roast them.[19]

Buffalo Bird Woman provided a rare insight into village life in the mid-1800s and the importance of corn, the technology developed around it, and the care of the crops and human beings.

Southwest

From 1540–42, Coronado depended on the Indigenous peoples of the southwest region for corn, beans, and squash, which were their staple foods. Coronado found corn from Mexico to Kansas. The Pima of Arizona, the Pueblo, and the Yuma, Cocopa, Maricopa, and Mohave of the lower Colorado River were skilled farmers who raised corn, beans, squash, and cotton.[20]

Corn was the staple crop for the Hopi. Because their land is dry they planted about a dozen seeds in a hill twelve to eighteen inches deep, which ensured some of the plants would survive. They spaced the planting hills about three feet apart. In the Hopi way, when a man married he moved to live with the wife's household. He was required to take with him seeds from his mother's family and plant them to prove he could raise crops. If he was successful, only then was he allowed to plant the seeds belonging to his wife's family.

Navaho planting of corn was a communal event in which corn was planted in "a spiral row which progressed clockwise toward the edge of the fields," although they also planted in regular rows. Corn was not hilled nor intercropped with squash, and beans were planted separately in hills. They dried and stored corn in pits.[21]

Alfonso Ortiz has written one the few available treatises illustrating the complex world view of the Tewa Pueblo and the interrelationship between world view, ceremony, and daily life.[22]

> The Tewa were living in *Sipofene* beneath Sandy Place Lake far to the north. The world under the lake was like this one, but it was dark. Supernaturals, men, and animals lived together at this time, and death was unknown. Among the supernaturals were the first mothers of all the Tewa, known as "Blue Corn Woman, near to summer," or the Summer mother, and "White Corn Maiden, near to ice," the Winter mother.
>
> These mothers asked one of the men present to go forth and explore the way by which the people might leave the lake. Three times the man refused, but on the fourth request he agreed. He went first to the north, but saw only mist and haze; then he went successively to the west, south, and east, but again saw only mist and haze. After each of these four ventures he reported to the corn mothers and the people that he had seen nothing, that the world above was still *ochu,* "green" or "unripe."

He travelled to the above world, where he was accepted, and returned to tell the people.

> The Hunt chief then took an ear of white corn, handed it to one of the other men, and said, "You are to lead and care for all of the people during the summer." To another man he handed another ear of white corn and told him, "You shall lead and care for the people during the winter."

Four days after the birth of a child, an intricate ceremony is conducted to name the newborn child.

> The naming mother holds the infant and the two ears of corn, while the assistant makes a sweeping inward motion over them with the hand broom, to gather in blessings for the child. The infant and the two ears of corn are proffered to the six directions . . .

White cornmeal is placed in the four corners of the Tewa home during this ceremony. The following prayer is said:

> Here is a child who has been given to us
> Let us bring him to manhood and womanhood
> You who are dawn youths and dawn maidens
> You who are winter spirits
> You who are summer spirits
> We have brought out a child that you may
> bring him to manhood and womanhood
> That you may give him life
> And not let him become alienated
> Take, therefore [proffering the child and the corn],
> dawn beings, winter spirits, summer spirits,
> Give him good fortune we ask of you.

The two ears of corn are put aside until the next planting season, when they are planted in the child's name.[23] Thus, we see that from birth, cosmology and corn are interwoven in the Tewa world view. In death, the person's "most prized and personal possessions" may be buried with them and for a woman this may include her "corn-grinding stone."

The world view of the Tewa equates women, corn, and clouds as symbols of fertility. The metaphor for corn is Mother. Ceremonies to honor corn are the green corn ceremony and the yellow corn dance.

> These dances celebrate and sing of corn maidens emerging from lakes to make their tremulo calls in response to the singing of youths. A third dance we call a cloud dance honors the primary colors of corn in a ritually prescribed order: blue, for the north, yellow for the west, red for the south, and white for the east. The women in this dance bring out ritually pre-scribed colored ears of corn in the order that places are named in prayers, chants, the order that dances are performed. So it is a nearly universal ritual prescription.

The men share a responsibility for the corn growing healthy by singing to the corn and thinking good thoughts. Ortiz notes that modern science has discovered that plants grow better with music.

Another metaphor in the Tewa world view is linking the emergence of the Tewa people from within the earth in four stages to four stages in growth of the corn plant: planting, when corn comes up, when leaves and cob form, and when it tassels.

They also think of each world level as representing a comple-
tion of a lesson, like learning how to overcome evil, learning
how to live together in groups, learning something else. The
humans have to learn something as a precondition to emerge
onto a next world level.[24]

Ortiz shares with us the detailed belief systems and intricate ceremonial cycle that provides a balance in the Tewa world. "This
entire cycle of works is tied to nature's basic rhythm and to the
Tewa's attempts to influence that rhythm for his well-being."[25]

Among the Tohono Awawtam, known as the Papago, songs are
sung for the corn when it is knee high, when the tassels form,
when the ears form, and when the corn becomes ripe. There are
also songs for the beans and squash, the "scared trinity of basic
crops in aboriginal America."

The Yaqui of Sonora have a corn origin story that begins
with a blackbird stealing corn and planting it for the people.
There is no rain and great adventures ensue to obtain rain until
finally the lowly toad is able to outwit the adversary and bring
the rain. In his analysis of world view and the role of corn Ortiz
concluded:

These are the four ways that I can discern, from a wide range
of reading, that Indian people have perceived and used their
traditions related to corn. They have seen it like the Papago
and Pueblo as mother, like Taraumara and Navajo people as
healer, and the Lakota as culture transformer. We scholars,
especially archeologists, know corn only as enabler, for it
enables a different kind of culture to arise. Or it enables
cultural achievement, which would not be possible without its
presence. Corn is an enabler for things to happen.[26]

There are many similar elements in the southwest corn culture
which relate the interconnectedness of corn, squash, and beans
within the world view and way of life of these cultures.

Central America

The antiquity of maize, as well as its importance, is attested
by the circumstances of its connection with religion, and its
acquirement of sacred characters. Centeotl, in Mexico was
goddess of maize, and hence of agriculture, and was known,

according to Clavigero, by the title, among others, of Tonacajohua, "she who sustains us."[27]

The "cultivation and domestication of plants *originated simultaneously* in several Mesoamerican areas." (emphasis added) The Tehuacan Valley has been cited as the place where domestication of corn originated around 5200 to 3400 B.C. "Independent agricultural development also occurred in Tamaulipas, on the northeastern periphery of Mesoamerica." Hurt utilizes *in situ* and evolutionary theories in explaining the origins of corn. Perhaps corn originated *in situ* in many areas, and perhaps it also was dispersed to other areas.[28]

The Mayan people live in the areas known today as Guatemala, Belize, and Honduras. Roderico Teni, a Mayan from Guatemala, shared his people's view of the role of corn in their lives:

> From the beginning of time, our Indian ancestors spoke to us about corn. In our sacred book that remains with us, the Popwuj, sacred book of the Maya, two of our original people, Hunahpu and Xbalanque, speak to their children about the raising of corn.[29]

The Maya have an all-night ceremony before selecting the field for planting. "In the spiritual context they request strength and protection for the being that is to be born, which is corn." Teni stresses the importance of everyone gathering to plant the field in one day. The host family provides a feast for the workers. The next day they go on to plant another family field. They clear the lands first and burn the field with the understanding that the ashes enrich the soil as well as ridding the soil of insects.

A seed ceremony is conducted to select the seed for planting. They use a planting stick, practice crop rotation, and allow the land to lie fallow. Teni explained the connection between fallowing the land and the birth of a child:

> By our custom, a woman, after giving birth, is made to rest for 30 to 40 days. This is done to allow her recovery and to insure that there are no problems in giving birth again in the future. The people feel the same about the earth and so are inclined to let it rest. During the time of rest from corn, the earth is sometimes planted with beans or squash, especially beans in order to help recover the nutrients.

Thus, we see the Mayan people have a clear understanding of the needs of the earth and human beings. In fact, Teni says that "corn is looked on (as) a living being, as a child might be." Planting time includes food as well as music and festivities.

The Mayan people have a harvest ceremony to give a thanksgiving for the corn. People are not to quarrel during this time. Corn is mixed with ashes and ground to make flour for tamales and tortillas. It is disrespectful to leave any corn in the fields, so they return to the fields to pick whatever might have been missed.

> This is done so that the connection that has been made is in no way disrupted—the spiritual connection to what they call the Heart of the Sky. It is seen as the need to make good on the request to the spirit to protect. If this is not done right—if there is waste of corn in the field—it is believed that the element of planting for that family begins to diminish and over the years the harvest will not be as good. The family itself will suffer because the meaning of corn, family and harvest is interrelated.

Teni discusses the relationship of Mayan astronomy and "spirituality that is inside of the corn cycle," and the Mayan calendar, which combine in the Mayan world view.

> This way of being for the Indian people is based on the knowledge that corn is alive, that corn feels and it cries. This is not easily lost and continues to this day. It's very important because life itself, culture and nutrition revolve around this food. It symbolizes the continuity and the importance of the cycle for the family—the maintenance of family, and, more than family, of the larger extended community.

The Mayan world view and way of life are interrelated and support each other within the ancient cultural values of the Maya.

South America

> In Peru the maize of Titiaca was considered sacred, and was distributed throughout the kingdom in small parcels to impart a portion of its sanctity to the granary wherein it was stored and in the garden of the Inca: "There was also a large field of maize, the grain they call quinua, pulses, and fruit trees with their fruit, all made of gold and silver."[30]

Corn was grown throughout the Andean valleys of south Peru, Bolivia, North Chile, Colombia, and Brazil. South America has a wide range of climates and altitudes, yet corn had adapted to the environmental ecosystems. At Tenochtitlan, Cortes (1518) cited fields of corn, and Pizarro (1533) in Peru "was frequently charmed with the beauty of the hillsides, clothed with this grain in all its stages, from the green and tender ear to the yellow ripeness of harvest."[31]

The Inca of Peru had a well-developed system around corn which included ceremonies for planting and harvest. At planting they asked the sun to protect the corn from excessive heat, and offered a drink made from maize into the rivers to ask for water for their fields. At harvest they "offered to the Earth corn and chica, and prayed for a good harvest. They threw corn into the river they were about to cross, or fish in, to propitiate its god." For feasts a bread was made of boiled maize.

Indigenous Agricultural Practices

Sauer tackles the question of the origin of maize from the wild grass Teosinte saying that the original "pure" corn preceded Teosinte. After quite a lengthy discussion of the issue, he concludes, "Present evidence points to a dissemination in all directions of the early forms from an unknown center."[32] "Grandfather Corn" or pod/husk corn may well be the ancestor or most ancient corn. Each kernel of the pod corn has its own husk and it appears to be a genetic seed bank for many varieties. Significant components of the agricultural cycle were found throughout many cultures whose staple crop is corn. These cultures generally employed slash and burn techniques to clear the land, and used digging sticks to break the ground for planting. They planted in hills and conducted two major weedings of the crop. Communal land ownership and rights of individual usage of the land, not individual ownership, were respected. The women were the genetic engineers who selected the best seed for planting, who kept the varieties of corn separate, and supervised the planting, as well as the agricultural methods. The agricultural practice of rotation of fields, rather than the crops, provides evidence that these peoples had a sound understanding of soil fertility. Watch towers were built to enable people to scare animals and birds away from the crops. Amazing irrigation systems were constructed where necessary. Underground pits and/or corn crib granaries were common methods of storing the crops. Very often the quantity of corn planted not only fed the people but supplied excess corn for trading purposes. This agricultural system provided a lifestyle in which people organized their society in permanent villages with systems of government, and ceremonial cycles. However, it is often difficult to ascertain the ceremonial cycle because past researchers did not necessarily connect world view and way of life, but focused on material culture. This discrepancy provides ample opportunity for further research. Each culture had some, but not necessarily all, of the above components. For example, corn might stored in pits, baskets, wooden containers, or granaries. But, overall, these components appear again and again as one reads about the many cultures in the western hemisphere.

Processing Corn with Alkali in the Americas

An intriguing study of the indigenous cultures of the Americas concluded "all societies that depend on maize as a major dietary staple practice alkali cooking techniques."[33] Katz discusses the scientific makeup of corn focusing on the proteins. Pellagra, a disease resulting from a deficiency of niacin in the diet, was found in places where people's only subsistence was corn and alkali processing techniques (wood ashes) were not practiced. Pellagra was common throughout the southern United States during the Depression and was found in South Africa and India. MesoAmerica, where alkali treatment of corn is common, experiences a low rate of pellagra. The scientific studies found that cooking with lime selectively enhances the nutritional quality of corn. Katz found that "cooking techniques in which alkali and heat are used clearly enhance the balance of essential amino acids and free the otherwise almost unavailable niacin."

Katz examined fifty-one cultures in the Human Relations Area Files, rating each by 1) the extent to which they grew corn, 2) the percentage of corn in the diet, and 3) whether they use alkali in preparing corn. The study found a high correlation between growing corn and the use of the alkali process in seven cultures. A deeper analysis of the data revealed that twenty-one of the cultures raise corn and use alkali processing; one culture traded for corn; seventeen raise corn but do not use ashes; and twelve did not raise corn. Of the seventeen that raise corn but do not use alkali processing, it was found that corn is not their staple food and they are likely to eat the smaller amount of corn they raise as fresh corn, which is boiled or roasted.

From the study of diverse cultures presented it is not difficult to understand that these indigenous peoples ate a widely diversified diet based on agriculture, gathering nuts and berries, fish, and meat obtained in hunting. Even the corn dishes were most often combined with beans, nuts, or meat, which adds protein to the diet. The fruits and berries added vitamins and minerals. The indigenous peoples of the Americas had a varied diet that provided all the essential nutrients for healthy living. This study, therefore, provides a broader perspective showing where corn and the alkali process are utilized throughout the Americas, but it could be more inclusive in cultures selected and provide a broader analysis of the diets of these cultures.

Corn has a cultural base in many diverse cultures in the western hemisphere. It occurs in the cosmology, in stories, ceremonial events, and agricultural practices within indigenous cultures. Belief systems surrounding the origins of corn vary from the Sky World of the Haudenosaunee to emergence from within the earth among the Zuni and Tewa Pueblo. Understanding the complex nature of indigenous cultures means recognizing the intermingling of world view and way of life. Researchers to date have done very little of this holistic research. The best examples of well-rounded world views are from the people within the culture, such as Parker, who was Seneca, on the Haudenosaunee, Ortiz, who is from Tewa Pueblo, and Teni, who is Mayan. These authors have a holistic view of their culture and present it in a manner that makes it understandable and meaningful.

> From its earliest appearance in written records, it was not simply an agricultural strategy or technology but a *cultural complex,* full blown, complete with stories, ceremonies, technology, customs, and etiquette. All indications lead one to believe that the Three Sisters complex is ancient . . . [34] (emphasis added)

The study of discrete elements extracted from a culture provides only a limited view of that culture. Discovering that people have ceremonies, tell stories, and sing for the corn suggests the degree to which corn is central to the culture.

Columbus and Dispersal of Corn to Europe, Asia, and Africa

Corn was evident from the very first contact between the two hemispheres. We do not know whether the first sighting of corn was in Haiti/Dominican Republic, or Cuba, but we do know corn

> was cultivated in Columbus's time from about 1000 km, about 600 miles, south of Santiago, Chile, up to Montreal in Canada.[35]

Columbus first encountered the Taino who greeted him with friendship, food, and gifts. The Taino had a thriving economy based on agriculture, gathering, fishing, and hunting. Las Casas saw vineyards of "three hundred leagues," and "game birds by the thousands," as well as foods such as yucca, dried fish, corn fields, and vast gardens of sweet yams.[36] They practiced rotation methods with their fields, and had irrigation systems where needed. There were estimated to be 100 or more Taino villages with 500–1,000 people. The Creation story of the Taino begins with four orphaned brothers who wander the sky islands and create this world. The Taino have ceremonies and dances in which they tell the Creation story, and oral traditions.

> Among the few Taino-Arawak customs that have survived the longest, the predominant ideas are that ancestors should be properly greeted by the living humans at prescribed times and that natural forces and the spirits behind each group of food and medicinal plants and useful animals should be appreciated in ceremony.[37]

An intriguing incident illustrates the different world views of the Taino and the Spanish. In 1494-95, Columbus demanded a tribute of gold from every Taino man, woman, and child. Of course, that much gold did not exist on these islands. The gold ornaments Columbus saw on his arrival had been with the people for generations. A thousand people with planting sticks gathered to offer the Spanish an alternative to the gold tribute. They said:

> We will feed you here on this island and also all of your people back in Castile. You don't even need to work.[38]

But the Spanish wanted gold and slaves, so they refused the offer. Apparently, using their agricultural skills to deal with more militaristic groups was sound diplomacy in this part of the world, but the Spanish had different values. This was a clash of two extremely different, even opposing, world views. The Taino people were devastated by European diseases and Spanish military policies, but they are not extinct.

Europe—Asia

Columbus took corn with him back to Europe and within forty years corn had spread throughout the Old World. Warman states the corn was the "product of the initiative of millions of people for thousands of years that produced a treasury of genetic knowledge."[39]

By 1525, in Europe, China, and Africa corn became the food of the poor people because it was cheaper than wheat. Because Europeans did not intercrop corn, beans, and squash, process the corn with ashes, they frequently suffered from pellagra. Among the wealthier Europeans, corn was used as animal feed, and did not become a significant part of their diet. "Corn did for the animal population of Europe what the potato did for the human population." The increase in domesticated animals meant more eggs, milk, butter, cheese, and meat thus providing more protein in the European diet.

Corn arrived in China around 1540 and was quickly adopted by the poorer people living in the hills because corn would grow where rice could not. Today, China is the "second largest producer of maize in the world, the first being the United States."[40]

Africa

It is generally believed that corn was brought to Africa by the Portuguese. The earliest account of corn was by Valentim Fernandes in 1502, and by 1550 there is no doubt that corn was raised in Africa. It is certain that corn was introduced to Africa very quickly after Columbus took corn back to Europe from the Americas. Miracle cites accounts of corn in Angola in 1617, Rhodesia 1561, and Zambia in 1798.[41]

Corn became the staple food to support the slave trade. The economic viability of corn was apparent in 1517, when the market for slaves for the New World increased to 100,000 slaves per year. American farmers raised corn for European forts, slave ships, and other nations. Corn, meat, and molasses became the diet for slaves, and "after slavery, maize became the staple of colonization, of colonial domination in Africa."[42]

Corn is grown in tropical Africa, the area between the Sahara Desert and South Africa. In some areas corn is the staple crop and in others it ranks within the top three staples along with manioc and bread or rice.

In looking for similarities in corn culture between Africa and the Haudenosaunee several areas emerged. Corn was incorporated into the culture of the peoples of Africa. An interview with Lawrence Mundia, from Zambia, provided the opportunity for a unique cultural exchange of information.

In 1990, when I came here, I went to Niagara Falls where there is a museum (The Turtle), and that's where I saw the corn for the first time. I read about it and said, Aha, so these people, they are our brothers and sisters. They have the same culture as us with the corn.

Mundia says corn was introduced to Africa by the early missionaries, but little research has been done on the history and sociological aspects of corn. The African people thought the corn came from Europe. "When the Europeans brought the corn, they never told us the truth. They always want the credit," says Mundia.

The Zambian people have three corn ceremonies: one for planting to be thankful for the rain, another when the corn begins to ripen (Green Corn), and a harvest ceremony. Mundia explained:

> There were so many stories, so many songs, dances. Because that's the way knowledge was passed from generation to generation. People didn't have schools. There were no books. So, to pass the knowledge about corn, how to plant corn, how to weed corn in the fields, how to harvest corn, how to store the corn, how to cook meals out of corn, all that knowledge was passed from one generation to another through story telling, and songs, dancing and so on.

When a ceremony was held, "each village would bring a different type of meal made out of corn." In the case of a death, the people from the same village provide food to the bereaved family. They have a system of clans that are named after animals, plants, and trees. These clans served to keep a balance in nature. The elephant clan was not to eat the elephant, therefore since each clan abstained from a certain food, the people did not overstress an element of nature, but kept a balance.

Many similarities were found in the material culture as well. Intercropping of corn, beans, pumpkins was standard planting practice as well as cooperative planting of the fields. They had both communal fields and smaller family gardens, and everyone helped with the work. The mortar and pestle is virtually the same in both cultures. However, one major difference found was that in Zambia corn is not put through the alkali process. It is fermented in water for several days, removed from the water, pounded in the mortar and made into a fine flour. Watchtowers are used to scare away birds and animals and corn is stored in granaries. An interesting aspect of planting included dual purposes for the corn. The women tended to plant the corn farther apart for quality, while the men tended to plant closer together for larger quantities for trade with other villages.

In Africa, corn is raised to feed people and has been adopted into their culture in ceremonies, songs, dances, and cultural practices. The similarity between the Haudenosaunee world view and way of life, and that of Zambia is quite remarkable. This happened because the corn was introduced to an indigenous people with a rich culture who accepted this corn with an attitude of thanksgiving and appreciation.

The Early Settlers:
Pilgrims, Squanto, and the Fur Trade

Corn was a vital factor at the time of contact between the first settlers and the indigenous peoples of the northeast. The fur trade was based on pre-existing trading practices with corn an important trade item.

In 1584, when Pennsylvania was first settled, the people saw "rich fields of maize," and the Native Peoples taught the Dutch and English arrivals how to plant corn.[43] It is widely known that the Pilgrims would not have survived without learning how to plant corn from Squanto. When the Pilgrims arrived they found corn pits and were grateful for the seed. Parker is quite emphatic in discussing the vital importance of corn in sustaining the early colonists:

> Few of us in these modern days realize the frightful struggles of these early pioneers to obtain food enough to sustain even the spark of life. It is recorded that some of the desperate Pilgrims, driven by the despair of hunger would even cut wood and fetch water for the Indians for a cap of corn. Others, we are told, "fell to plaine stealing both night & day from ye Indeans of which they (the Indians) greviously complained."[44]

According to Bradford's, *History Plymouth Plantation,* Squanto taught the Pilgrims how to plant corn and put fish in the hill for fertilizer. What hasn't reached the textbooks is the background on Squanto. He, and other Indians, had been kidnapped in 1614 by Captain Thomas Hunt, "master of a vessel with Captain John Smith," and taken to Spain to be sold as slaves. He was freed by an Englishman and spent seven years in Europe. He returned in 1619 as a guide, to find that his people had been wiped out by an epidemic. It was in 1621 that Samoset and Squanto taught the pilgrims to plant corn. Squanto also showed the Pilgrims how to trade with local Indians for furs to acquire supplies they sorely needed. In 1622, Squanto fell sick and died.[45]

Hurt and Ceci maintain that Squanto learned about using fish as fertilizer during his stay in England and that this technique was not native to New England's indigenous peoples. The basis for Hurt's conclusion is that only Bradford mentions the fish.[46] Fertilizing with fish is said to date back to the Romans. Ceci's argument is interesting but the undertone maintains an ethnocentric attitude common to Western civilization that only the colonial settlers were intelligent enough to conceive of the idea of using fish to fertilize the planting hills.[47]

The early settlers adopted much of the agricultural and cultural knowledge of the indigenous peoples they encountered. In Connecticut's history the same pattern of the first colonists learning how to plant corn can be found. Captain John Smith, in writing Virginia's history, states: "Such was the weakness of this poor common-wealth, as had not the salvages fed us we directlie had starved."

> And thus it is that the maize plant was the bridge over which English civilization crept, tremblingly and uncertainly, at first, then boldly and surely to a foothold and a permanent occupation of America.[48]

Corn quickly became legal tender in the early colonies. In 1621 it sold for "2s. 6d., or 62 cents per bushel." A patent was placed on corn in Massachusetts Bay during the 1630s forbidding colonists to sell corn without the patent. Corn was used to pay taxes throughout the colonies, and if a cow damaged the corn crop during the night, the owner was liable for the damages. Wages were paid in corn for servants and workmen. When there was not enough corn, people were forbidden to take any corn out of their area. The same value of corn was present in New Netherlands, Rhode Island, Maine, and New Hampshire. William Penn, Governor of Pennsylvania, declared in 1683 that the "seal of Kent county should be three ears of Indian corn."[49] Corn was exported and imported among the colonies in the mid-1700s. *Lawson's History of North Carolina* (1714), describes a land of plenty, a virtual paradise, in which without corn "it would have proved very difficult to have settled some of the Plantations in America."[50]

Corn had a significant impact during the fur trade era. The Huron and Haudenosaunee already had a trading empire with Indians living inland. They traded corn and tobacco for furs. As Indians became more dependent on European trade goods the importance of these pre-existing trade routes expanded. In 1625 Bradford "sent a boatload of corn up the Kennebec River to trade with the interior tribes for furs." Descriptions of intertribal warfare centered around the battle to dominate the fur trade. The far-reaching boundaries of the fur trade extended to the Midwest and Great Plains where "the agricultural tribes acted as middlemen between white traders and nomadic hunters." Wessel states "agricultural products lubricated the system" of the fur trade.[51] The Van den Bogaert description of Mohawk and Oneida villages cited earlier in this chapter was obtained on his journey to negotiate with the Haudenosaunee over fur trade prices. The French had offered better prices and Van den Bogaert was trying to form a trade alliance in which the Haudenosaunee would trade exclusively with the Dutch. The Haudenosaunee were often caught in this three-way triangle of French, Dutch, and British traders. Thus, learning and practicing American Indian agriculture not only enabled the first settlers to survive, it became their monetary system and undergirded trade systems.

Pehr Kalm, a Swedish botanist, came to North America between 1748 and 1751 to study American plants.[52] He describes seeing miles of nothing but maize in his travels. Kalm follows this with a description of planting methods that sound suspiciously like the hill planting with fish as fertilizer. The settlers plowed their fields in such a manner as to create furrows or hills. They did not intercrop corn, beans, and squash, but planted the corn in hills. They soaked the corn seed before planting and were cautious in seed selection. The first weeding, or plowing, pushed the soil up around the corn creating another hilling process. Corn was stored in corn cribs, and mortars were used to pound the corn.

Kalm shares an interesting story about Cadwallader Colden, who was collecting the many varieties of corn seed. He labelled one *"Zea semine nudo,* a species which does not have the hard membrane." As it turned out, "he had been deceived," because the corn without the membrane had been processed with lye. He planted this corn and of course it did not sprout. There must have been some chuckles among the Indians who gave him that corn.

Indian corn was utilized for human food and for animal feed. They tended to mix rye and corn flour for bread. The English would cook pumpkin and mix this with the maize flour. Kalm describes the time he spent with the Dutch who for long periods of time had only maize mush and milk to eat.

Cracked corn or hominy was made by the settlers and called grits. This corn was placed in a mortar with a little water and pounded with a wooden pestle to loosen the hulls. It was then cooked with water and meat was added. Kalm relates that the settlers also used a lye solution to loosen the hulls. They made beer from their maize crop.

Corn was a major source of food for the domesticated animals. Each farmer had a mortar that was used just for pounding corn for animal feed. Thus, we find that many aspects of American Indian agriculture from soaking seeds to planting in hills to using lye to remove the hulls had been adopted by the European settlers.

> Agriculture in the first two hundred years of American history was woodlands farming . . . Girdling trees and planting corn after grubbing out the underbrush dominated western agriculture until well into the nineteenth century. In short, white farmers continued to practice what they knew best, the agriculture of the woodlands Indians.[53]

Another perspective on raising corn and its effects on the European settlers is presented by Crosby, who cited a direct relationship between land, corn, and health.

> In 1751 a cocky Benjamin Franklin wrote that for every marriage per hundred people in Europe, there were two in British America, and for for every four births per marriage in Europe, eight in America. He estimated that the population of the American Britons was doubling every twenty years.[54]

Research figures indicate that the fertility rate around "1800 was about twice the highest rate of the baby boom in the United States in the 1950s." Crosby theorized this tremendous population increase was possible because they had good land, corn was amazingly productive, and they had adopted American Indian planting techniques, which they maintained for two and a half centuries. The Europeans learned that European farming techniques, such as broadcast sowing of wheat seed, did not work in the Americas. Crosby cites Maryland where even the "aristocracy ate hardly any other bread." William Cobbett, an Englishman who "fled" to America, was a "radical and agricultural expert" who said the Thirteen Colonies would not have survived "without Indian corn." "Maize was, he proclaimed, not only the greatest blessing of the United States, but "the greatest blessing God ever gave to man."

Crosby's study maintains that population increase and prosperity depend on an ample food supply which Indian corn provided. The settlers had large families which they needed to work the land, and they had "more men to fight the dispossessed Amerindians;

and the open spaces kindled ambitions for dominion." The gift of Indian corn and Indian agricultural knowledge enabled the survival and health of the early settlers, but it also dispossessed American Indians from their lands. Although the exchange of corn and knowledge was evident, there were also great hardships and wars during these early days.

Military Policies to Destroy Haudenosaunee Corn

Denonville—The French Attack on Corn

There were two major military efforts to destroy the military strength of the Haudenosaunee by destroying their corn supply. The first was in 1687, when the French commander, Denonville, attacked the Seneca village of Ganondagan. The French and English were waging war over territory and the fur trade. Each side was amassing "Indian allies." Greenhalgh's description of the Seneca villages in 1677 documented large villages with tremendous quantities of corn. It was these four Seneca villages (see Figure 6.2) and their fields which the French Marquis de Denonville set out to destroy. Marshall cites three reasons for the attack: 1) to destroy Seneca villages and fields to gain control over them; 2) to build a fort at Niagara; and 3) to insure French control of the fur trade.[55] However, Denonville's journal cites different reasons: 1) the Iroquois were feared by everyone; 2) the Iroquois did not accept Christianity; 3) to gain control of the fur trade; and 4) to settle disputes with the English over territory.[56]

The four Seneca villages contained a total of 324 houses and an estimated population of 14–15,000 people. There was ample evidence that the Seneca were a well-settled people who raised huge fields of corn, hogs, and stored their corn in underground pits as well as in a large palisaded granary on Fort Hill.[57] The largest village was Ganondagen, located near Victor, New York, which has been declared a historic site by New York State. Marshall states of the Iroquois:

> At this period, the northern, middle and western parts of the state of New York were a howling wilderness, and the Five Nations ranged their hunting grounds in unmolested freedom.[58]

Yet, the evidence is quite the contrary. These were large settled villages as Greenhalgh documented in 1677 and as Denonville found in 1687.

Marshall, in researching the sites of these Seneca villages, had difficulty locating the exact site and turned to oral histories to find evidence. Ganondagen is called the "ancient village of the Senecas" by John Blacksmith in one of the oral histories. This village was well known as "Gaosaehgaaah or the basswood bark lies there," because the Seneca had constructed troughs to a spring, the one major source of water. Once the site at Victor had been identified, Marshall found oral histories handed down from the first settlers of Victor, which included the battle between the French and Senecas. Evidence of a former Indian village at Victor included "Indian hatchets, gun locks of rude construction, gun barrels, beads, pieces of brass kettles, stone pipes." In fact, so much iron was found that it supplied the blacksmith shop. "Thousands of graves were then to be seen, many of which are yet visible, and rude implements, of savage construction, are often found on opening them." There was evidence of interaction with European settlers and acquisition of iron and brass, but apparently there were no laws forbidding intrusion into Indian graves.[59]

> There is a tradition among the Seneca that Denonville at-
> tacked while a major portion of the Seneca warriors were
> away somewhere to the west.[60]

Marshall provides two oral histories from Senecas relating to the attack by Denonville that confirm the warriors were away. The four villages Denonville destroyed had been deserted and burned by the Seneca as they fled for their lives. Only 800 Seneca men attempted to ambush Denonville's vast army of French settlers and militia, and their allies the Christian Indians (Mohawks), Algonquins, and Ottawa. The French army spent ten days, from July 14–24, destroying new fields of corn and corn stored in pits and the palisaded granary. Denonville's journals state they destroyed 1.2 million bushels of corn as well as beans and other vegetables. The Indian allies of the French thought the destruction of corn as not worthwhile and Denonville had a difficult time keeping them with the army.[61]

Documentation reveals this method of destroying corn to control Indians was not a new strategy. Sir Richard Grenville, in a dispute over a stolen silver cup, ordered an Indian village and corn fields burned in July 1585, at Roanoke, Virginia.[62] The English destroyed more than 200 acres of Indian corn during the Pequot War in 1637, and the Puritans "took possession of one thousand acres of corn, which was harvested by the English," during King Philip's war in 1675.[63]

The Seneca land in the Genessee Valley in 1687 was certainly not a howling wilderness, but a land of many well-settled agricultural villages, as evidenced by large quantities of corn.

The American Revolution—Clinton and Sullivan

The journals of the Clinton and Sullivan Expedition describe the Iroquois lifestyle in 1779, nearly a hundred years after the Denonville attack. The descriptions included village size, style of houses, the quantity of corn, beans, squash, fruit orchards, and domesticated animals. This data illustrates cultural continuity, as well as cultural interaction and adaptation.

All accounts of wars, in every historical time period, are full of atrocities and descriptions of man's inhumanity to man. The American Revolution and the Clinton/Sullivan campaign contain these same atrocities and background information provided here serves the purpose of dispelling misinformation.[64]

The commonly accepted view of the role of the Iroquois in the American Revolution presented in textbooks states that because the Iroquois sided with the British they lost their lands. The historical documents provide evidence that the Iroquois tried to remain neutral. In separate meetings with both the British leaders and the American rebels the Iroquois maintained their position of neutrality. In July 1775 the British met with the Iroquois to request that they remain neutral in this disagreement. The American rebels met with the Iroquois in August 1775 with the same request. In both instances the war was explained as a family quarrel between a father (the King of England) and children (the American colonies). A year later, a meeting was held at German Flats, on August 6, 1776, in which the Haudenosaunee renewed their pledge of neutrality to Philip Schuyler, saying, "This then is the Determination of the Six Nations—Not to take any part, but as it is a Family Quarrel to sit still and see you fight it out."[65] The British were wealthier and continued providing the flow of vital trade goods with the Iroquois. The American colonials were viewed as land grabbing people. In August 1777, an unforeseen event broke the neutrality and rallied the Seneca to join the American Revolution on the side of the British.

> To create a favorable impression, as allies of the British, the Seneca were invited with their tribal brothers to witness the battle at the taking of Fort Stanwix. It was an invitation to

Sky Woman

(All Ernest Smith prints in this section courtesy Rochester Museum of
Arts and Sciences, Rochester, NY)

Council with Tadadaho

Three Sisters

Picking Wild Strawberries

Maple Sugar Time

Sullivan's Campaign

witness a holiday party, in which the colonists should be whipped by the well-drilled and handsomely dressed British miltia. This would impress the Iroquois with power of the British and the expediency of supporting them against the rebellious settlers! All that the Indians were asked to do was "to sit down, smoke their pipes and look on," said Mary Jemison. She continued her account by saying that, "Contrary to expectations, instead of smoking and looking on, they were obliged to fight for their lives; and in the end of the battle were completely beaten, with a great loss in killed and wounded. Our Indians alone had thirty-six killed and a great number wounded." Such was the participation of the Seneca in the first real battle, that of Oriskany, *the place of nettles*.[66]

The Seneca then joined the war on the side of the British to avenge the loss of Seneca people, including women and children, at Oriskany. The Mohawks had always been more aligned with the British under Joseph Brant. They attacked the colonial towns of Cherry Valley and Wyoming. Parker goes on to say,

It must be said, however, in justice to them, that no women or children were killed by direct attack. Smarting as they were from their former defeat, the Seneca at Wyoming had full opportunity to glut their revenge, and yet the restraint that they showed is no less than remarkable. Every historian knows that this battle might have been a massacre indeed, had the Seneca so willed it. They had not forgotten the murder of their own women and children.[67]

Colonel Goose Van Schaick of the American army attacked and destroyed three Onondaga villages in April 1779, which broke Onondaga neutrality.[68]

There were divided loyalties within each of the Iroquois Nations, with supporters for the British and the colonists, and those who tried to remain neutral. After the Oriskany battle and the Onondaga attack, the Mohawk, Seneca, Cayuga, and Onondaga fought on the side of the British.

The Tuscaroras had tried to remain neutral and many of the Oneidas were influenced by their New England (and pro-rebellion) missionary Samuel Kirkland and joined the colonials. The divided loyalties among the Oneidas is clearly illustrated with the incident between two Oneida brothers. One was on the British side, the other on the American side. The one on the American side was captured by the British and Seneca. The brother on the British side chastised his brother for joining the wrong side, but would not kill him. He turned his brother over to the Seneca, who promptly dispatched him.[69] This war, as is true in all wars, saw people with divided loyalties and there was not a unified position among the Haudenosaunee, either on a family, Nation, or Confederacy-wide basis.

On February 25, 1779, Congress authorized the Commander-in-Chief, George Washington, to destroy the Iroquois villages and crops. Washington was unsuccessful in getting General Gates to lead the expedition, but selected Sullivan a thirty-nine-year-old lawyer from New Hampshire who was a Major-General in the American army. Washington's letter to General Gates, which was forwarded to Sullivan, stated:

> The object will be effectually to chastise and intimidate the hostile nations; to cut off their settlements, destroy their next year's crop, and do them every other mischief which time and circumstances will permit.

Washington wrote to Sullivan:

> The country must not merely be overrun, but destroyed . . . You will listen to no overture of peace before the total ruin of their settlements is effected.[70]

Seneca and Cayuga Villages in 1779

Descriptions of the Seneca villages and the fertile Genesee Valley provide a picture of the Seneca as having maintained their own lifestyle and culture as well as having incorporated tools and plants from the settlers. The villages were no longer surrounded by palisades and many homes were now constructed of logs and boards, and painted, which shows the change from longhouses to log homes. Parker says, "The Seneca felt themselves secure, and had it not been for the white man's war they would have been happy indeed."[71]

Sturtevant's description of the Seneca when Gen. Sullivan invaded their territory includes finding "the lands cultivated, yielding abundant corn, extensive orchards, and a regularity in the arrangements of their houses which announced prosperity and enjoyment of property." General Sullivan found the Seneca had grown "every kind of vegetable that could be perceived," and another record catalogues "corn, beans, peas, squashes, potatoes, onions, turnips, cabbages, cucumbers, watermelons, carrots and parsnips."[72]

The journals of officers in the Sullivan expedition are filled with descriptions of the towns/villages they found throughout the Susquehanna River region, along Seneca Lake, Genesee Valley, and the Allegany River. Villages were called towns in the journals and varied from as small as three houses to as large as 128 houses at the Great Genesee Castle. There were many villages with an average of fifteen to twenty houses. In his journal, Lieutenant Erkuries Beatty said:

> This town was one of the Neatest of the Indian towns on the Susquehanna, it was built on each side of the River with good Log houses with Stone Chimneys and glass windows it likewise had a Church, burying ground and a great number of apple trees . . . we burnt their town to ashes . . .[73]

The Indians in this town had obviously been Christianized, but the colonial army destroyed the town anyway.

The army consistently described the areas around the villages as cornfields usually of 100 acres or more. The colonial army recorded the number of houses in each village, the name of the village, the amount and variety of vegetables found, and the amount of corn. In every village, they burned the entire village and destroyed the fields of corn. Lieutenant Beatty reports on August 30th:

> Our Brigade Destroyed about 150 acres of the best corn that I Ever saw (some of the Stalks grew 16 feet high) besides great Quantities of Beans, Potatoes, Pumpkins, Cucumbers, Squash & Watermellons, and the Enemy looking at us from the hills did not fire on us.[74]

Major Burrowes described a village six miles from the Genesee Castle on September 15th:

> The whole army employed till 11 o'clock destroying corn, there being the greatest quantity destroyed at this town than any of the former. It is judged that we have burnt and destroyed about sixty thousand bushels of corn and two or three thousand of beans on this expedition.[75]

The journals provide descriptions of fruit orchards over fifty years old, and one fruit orchard with 1,500 trees. A long-established permanent settlement and sound agricultural practices were in place to provide an orchard of 1,500 trees.

Each of the forty-four destroyed villages was abandoned before the colonial army arrived because the Cayuga and Seneca were vastly outnumbered by the colonial army. Beatty reports the army burned an abandoned Scottish settlement and the Glasford house because these non-Indians had gone over to the British side.

Corn had another vital role in the Sullivan expedition—that of feeding the army. Major Burrowes said on September 3, 1779, "our-selves suffer some hungry hours, for not being near any cornfields." The colonial army had to wade through swamps, over mountains, and endure endless days of rain and sickness on half rations, which they had to carry along with their baggage. The men were happy to find the Iroquois towns knowing there would be corn and they would eat well that day. "We eat meat twice in three days, and bread once in four or five days. The Country abounds with corn and beans which we solely live on. Salt very scarce." Lieutenant Beatty reports on August 27th, "Marched this day six miles within two miles of Shomony, where they had planted a great deal of corn beans &c which we feasted heartily on." An interesting observation was made on September 3rd by Beatty on finding a town on the east side of Seneca Lake, where the corn left cooking over the fire was purple. The army of Clinton and Sullivan swept through the Cayuga and Seneca villages, feasting on the vast quantities of food and destroying the rest.[76]

Parker describes the Genesee Valley as the fertile garden region of the Seneca in which Sullivan saw, "the land of the savages to be, not a tangled wilderness but a smiling blooming valley, and the savages domiciled in permanent houses and settled in towns."[77] William Stone, in his 1838 book *The Life of Joseph Brant*, states that they found farms in the Indian country "resembling much more the orchards, and farms, and gardens of civilized life."[78]

Sullivan's final report on September 30, 1779, stated:

> The number of town destroyed by this army amounted to 40 besides scattering houses. The quantity of corn destroyed, at a moderate computation, must amount to 160,000 bushels, with a vast quantity of vegetables of every kind . . . I flatter myself that the orders with which I was entrusted are fully executed, as we have not left a single settlement or a field of corn in the country of the Five Nations . . . [79]

Dennonville's destruction of 1.2 million bushels of corn, and Sullivan's destruction of more than forty villages and 160,000 bushels of corn illustrate the major role corn held in the strength of the Haudenosaunee. The detailed descriptions of vast fields of corn, fruit orchards, and other vegetables clearly illustrate the continuity and sophistication of Haudenosaunee agricultural practices. The mention of purple corn indicates that varieties of corn were grown and utilized in the diet. The vast quantities of corn, fruit, and vegetables on which the colonial army feasted are indicators of a healthy Iroquois economy.

Adaptation is evidenced by the change in housing from bark covered longhouses to houses built of hewn logs typical of that time period. The colonial army certainly saw that the Haudenosaunee were not savages living in a howling wilderness. Cultural interaction obviously took place through the well-established diplomatic networks that were vital in maintaining the Iroquois neutrality in 1775–77.

There is no doubt that the Sullivan expedition destroyed the women's economic base for the year 1779, but crops were replanted the next year. All the Seneca and Cayuga villages were not destroyed as Sullivan thought, and some of the Haudenosaunee returned to those remaining villages. However, the bulk of the Seneca and Cayuga fled to the British Fort Niagara after the Sullivan expedition. The British did not have the provisions to feed them and it was a harsh winter. The greater disaster was the loss of land that came after the American Revolution. Corn could be planted the following year, but the loss of lands was forever.

After the American Revolution, the Haudenosaunee lost vast pieces of land through unscrupulous land company dealings and treaties, both federal and state. They were in a state of turmoil and distrust over these land transactions and trying to adjust to the system of small land bases called reservations, which were established in treaties. Haudenosaunee economy was devastated by the loss of land for both hunting and agriculture.

The Quakers

In 1789, Secretary of War Henry Knox decided that war or conquest was too expensive and changed the policy to "civilizing" the Indians, hoping for the same end result, "acquisition of Indian land."[80] The philanthropic Quakers quickly became the agents for leading the Indians from "barbarism" to so-called "civilization." The Quaker policy was to train the men in agriculture and the worth of private property and train the women to be in the house. Of course, these policies were the opposite of balanced economic roles in Haudenosaunee life where the women were in charge of agriculture while the men hunted, fished, traded, engaged in diplomacy and defense of their lands.

In May 1798, the Quakers arrived among the Allegany Seneca to establish a model farm with fenced fields, domestic animals, plows, wheat, and oats. The boat with tools and equipment had not yet arrived so the Seneca women gave them corn, beans, squash seeds. Wallace describes the initial contact quite differently saying the Quakers found "about four hundred hungry people" because floods had damaged the corn crop, and that their first view of the Seneca "was a band of drunken Indians at the land company store at Warren," Pennsylvania.[81]

The Seneca met in council several times to consider the Quakers' proposal to build a model farm, establish a grist mill, and teach the women to spin. The Seneca stated they had deliberated over the matter and allowed the Quakers to stay. It was a time of turmoil for the Seneca with all the changes being forced on them, including internal upheaval over how to deal with these changes.

The Seneca women's response to the Quakers was "levity" while the young men were distrustful, having been so recently "cheated out of their lands." As with any society, there was behind the scenes political maneuvering between those who favored the Quaker plan, the progressives, and those who did not, the conservatives. In the midst of all these council meetings to evaluate the situation and make decisions, came the visions of Handsome Lake, which, for the most part, did not disagree with the Quaker plan. Cornplanter heartily approved of the Quaker efforts. Wallace provides a detailed description of the interaction of the Quakers, Handsome Lake, Cornplanter, and Red Jacket, which is replete with political intrigue.[82]

The Seneca economy changed over a short time period from the fur trade, which ended in 1808, to lumber and wood products, to wage labor on railroads and construction. In terms of agriculture, corn remained the staple crop with the women raising sufficient quantities for daily use, trading, and the ceremonial cycle. The form of agriculture accepted by the men was to raise domestic animals, wheat, and oats, but these changes were not added to the ceremonial cycle.

An agricultural experiment was finally agreed upon in 1801. The corn field was planted in the traditional manner, but every other row was plowed. The plowed rows showed a higher yield so, according to the Quakers, the plowing became more common.[83] Oral tradition among the Seneca provides another version of that ex-

periment saying that for the first few years the plowed land produced more corn, but after a period of time the productivity de-
creased and the people returned to hill planting in the traditional
manner.[84] Parker, writing in 1910, describes the hill planting system, which continued in his day. However, the young men found
they could earn money plowing for others and thus began the tran-
sition to wage labor or "cash-producing activity."[85]

The Seneca, while flexible and receptive to new ideas, were
also cautious and "selectively adopted" those new ideas that best fit
their traditional customs, and corn continued to be the staple crop,
and was sold to white settlers. The women incorporated soap
making, spinning, knitting, and weaving into their wintertime
activities and continued their vital role in agriculture. The men
were actively engaged in the lumber industry and preferred that to
farming.

Seneca women held a high status in their nation. They owned
the lands, the crops, and had control over the food supply. They
made political decisions and selected and could impeach the chiefs.
They could veto war by withholding food. Brown states that in an
agricultural society the natural resources are land, seeds, and tools.[86]
Since the women controlled these three items it is safe to assume
they had a significant say in whether, and to what extent, the
Quaker policies were adopted.

When the Seneca resisted the changes in men's and women's
roles in private property, in plowing, and so on, they were being
selective and making rational decisions based on their culture. One
major area where the Quakers did not succeed in making a change
was in the concept of private property. If they had been successful,
the Haudenosaunee would not have a land base today. The people
actively refused the alien concept of private property because it
could not be balanced with Haudenosaunee world view. The adaptation that did take place was to build frame houses with glass
windows and to live "in a more dispersed settlement pattern."[87]
With Handsome Lake's approval, the small nuclear family and
homestead established a significant change. Even though Seneca
people were beginning to establish nuclear family homesteads they
did not incorporate the concept of private property. The land was
used by individual families, but the land *belonged* to the Seneca.

The educational and technological customs of the whites were
to be integrated into a Seneca society that retained as a social
value the village communalism, reserved land title to the nation
itself, and depended upon traditional reciprocal gift-giving

rather than commercial sale as the mechanism of internal distribution. . . . Thus the society envisioned by the prophet (Handsome Lake) was to be an autonomous reservation community, using white technology, but retaining its Indian identity.[88]

The Seneca people based their decisions on selectively adopting those elements of white society which they saw as good, and rejecting the harmful elements such as drinking and gambling. They took more of a "wait and see" attitude toward school.

Overall, the men had always helped clear the fields and participated in the harvest and they continued to do so. The Quakers did get the men to raise domestic animals and plant hay and oats. The women continued the corn agriculture and incorporated the Quaker household skills into their winter activities. The change to nuclear families living on homesteads was difficult and only happened gradually over a long period of time.

Selected adaptation of elements from the white world were incorporated into the Seneca way of life. Ideas that could not be blended into the existing world view, such as private property, were not accepted and have not been accepted to the present day. The Haudenosaunee were cautious, conservative, and careful about new ideas and the effects of those changes to the seventh generation, a philosophy that continues to be in effect.

The Role of Corn in Building the United States

When the pioneers set forth to acquire land and fortunes in America, "tomahawk rights" and "corn titles" were the unwritten law of the land. When a pioneer selected a piece of land he would "blaze his name, the date and a number of acres on a tree," which established his "tomahawk right" to the land. When he settled the land and planted his first crop of corn he had "corn title" to the land.[89] "Corn patch and cabin rights" established ownership during the Revolutionary period.[90] In 1776, Virginia passed a land ownership law that required living for one year on a piece of land, planting a corn crop and building a cabin.

Indian corn and farming practices were thoroughly integrated into the small homesteads of settlers including identical methods of seed selection, determining when to plant, planting in hills, harvesting, husking, and storing corn. Abraham Lincoln said that his job, as a young boy, was to plant two pumpkin seeds in every other hill.[91]

"This generation is very sure to plant corn and beans each year precisely as the Indians did centuries ago and taught the settlers to do, as if there were a fate in it," wrote Thoreau in the late 1840s.[92]

Whether hill planting or plowing provided the greatest productivity remained a debate into the early 1900s. Enfield (1866) described planting corn in hills four to five feet apart. Enfield favored drill or row planting because he felt it was more productive than hill planting. In a comparison of productivity of the two methods, twenty-one different planting arrangements were analyzed. The independent variable, the distance between hills or rows, ranged from twenty-four to forty-eight inches. Analysis showed greater productivity when corn was planted twenty-four inches apart, whether in hills or rows, and overall larger quantities from rows planted twenty-four inches apart.[93] Plumb cites experiments at the New York agricultural station that showed that planting corn in hills of four or five stalks gave the most satisfactory yields.[94]

There exists in the literature a sense of nostalgia for pioneer settlers, and later for the homesteaders who established homesteads in the west. Hardeman, writing in 1981, bemoans the loss of the "old corn culture" as having "slipped away so quietly" to become an "all-but-forgotten era." He acknowledges the "heritage of Indian, European, and African America is inseparable from maize or Indian corn."[95]

Hardeman also provides a less picturesque view of farming when he explains that many times the pioneer men were off in military expeditions or "Indian campaigns" or "political assemblies, trading, salt boiling, hunting, trapping, scouting for new home sites, or else taken off the work force by illness, injury or death." In such instances it was the women and children who kept up the tremendous demands of farm "drudgery." Yet, it is only is this context, that of women and children doing the work, that farming is drudgery. Otherwise it is hard work which built strong character in the Jeffersonian ideal of "the spirit of true democracy and virtue" of those tillers of the soil. The hard work on American farms taught self-reliance, teamwork, the discipline of work, and provided a sense of personal worth from being needed, and quite necessary to support the family. By the 1840s people recognized that working on a farm builds character and felt that poverty in the city did not build the proper American character.

The national identity incorporated corn as a design in art, architecture, and symbology. In 1803, during Jefferson's presidency, the design at the top of columns on the Capitol was a corn motif. Corn was used to decorate the Sioux City corn palace, built in 1887. In 1892, the four hundredth anniversary of Columbus, corn was adopted as the national flower of the United States.[96]

In 1917, Jeannette Young Norton, in her effort to support the war effort for World War I, urged the American housewife to:

> EAT CORN BREAD! . . . at least once a day, to release most of our wheat crop for the sustenance of the Allied Armies.

This plea supported a "corn message" sent out by Congress in 1917:

> Corn is America's biggest crop, and I appeal to my sister women to rise to a loyal patriotism and use it in every way available. Our reputation as housewives and mothers is at stake.

Murphy exhorts Americans to grow more corn, and respect this "prolific grain," which was "the source of so much wealth and power to the nation," and felt this respect for "America's national food" would "cause a quicker flow of patriotic feeling in the breasts of my countrymen."[97]

American Indians had many varieties of corn which they knew how to keep pure, and they knew how to cross-fertilize to develop new strains of hardier plants. From these varieties of Indian corn, the American farmers cross-fertilized to produce additional varieties. In 1866, Enfield lists a total of thirty-four varieties. Plumb (1895) lists a total of forty-six varieties, which is an increase of twelve varieties over twenty-nine years. The names of some of the varieties are obviously Indian names, such as King Phillip, named after the chief of the Wampanoags; Waushakum, a mixture of Canada and New England Eight Rowed types; Narragansett; and names of European descent named after people and places, such as Longfellow which was developed in Massachusetts, Rhode Island White Flint, Chester Co. Mammoth, and "Blount's Prolific after Prof. A. E. Blount in Tennessee." Enfield says these varieties are hybrids, which he defines as cross-fertilization of two varieties. He provides directions for selection of seed to produce the best qualities for the climate.[98]

In 1920, none of the corn planted in the corn belt was hybrid, but by 1956 hybrid seed was used to plant 99.9 percent of the corn crop. The definition of hybrid means a double cross hybrid, that is A and B are crossed to produce AB. Then, C and D are crossed to produce CD. The crossing of AB with CD produces the double cross

Figure 7.1. Double Cross Hybrid Seed

hybrid ABCD.[99] In 1925, double cross hybrid seed developed into a booming business for seed companies.[99]

Why was there such an increase in varieties? Why the change to 99.9 percent hybrid corns? The push came for uniformity to enable efficient use of machinery, and to increase productivity per acre of land which would in turn provide increased production of corn products.

Enfield, writing at the end of the Civil War, vigorously urges farmers to improve their productivity by using better seed selection methods and working harder. Farmers are viewed as a "large and respectable body of men" and

> the continued existence of our population is literally suspended upon the tillage of the earth. The farmer feeds the community, and every member of it thus daily, and almost hourly, reminded of his value and importance in the social scale.[100]

After praising the worth of farmers he turns to berating them for only producing thirty-three bushels per acre when, with more industry, they could be producing 100 bushels per acre. He discusses the importance of farmer's clubs and farming journals, and experiments, but seems to know that many of these farmers can't read very well. He encourages farmers to increase the crop by 1870 to "feed not only our own people, but half the population of Europe," and says that in three years of increased productivity they could "extinguish our national debt, and leave a balance in the treasury."

The corn crop, in 1885, was 86 percent larger and worth more in dollars than the total for all other grains produced in the United States. The abundance of corn moved Furnas to declare "Corn is King" when he was appointed to represent the "young agricultural giant, Nebraska, at the 'World's Industrial and Cotton Centennial Exposition,' New Orleans, La., 1884–5." The corn yield in Nebraska averaged forty bushels per acre but he was not satisfied and urged a yield of seventy-five bushels.[101]

Murphy extrapolates detailed statistics on corn productivity, citing the 1888 crop of 2.2 billion bushels of corn. He praises the value of corn as the most adaptive to soils, climates, and purposes. In comparing corn to wheat, he says:

> [P]ound for pound and bushel for bushel, it supplies as much nutriment as wheat itself... though, on account of its abundance, costing considerably less than one half as much. Corn, taking a series of years together, is far more certain than wheat.

Murphy's 1889 lecture before the National Agricultural Society at the International Congress of Millers, Paris, was designed to encourage Europeans to include corn in their daily diet. He lectured on the merit of corn productivity, adaptability, and uses for corn as food for human beings.

Murphy uses several math and geography examples to explain the vast quantities of corn produced in the United States. He cites the "Corn Patch" as the six-state area of Indiana, Illinois, Iowa, Missouri, Kansas, and Nebraska, which raised, in 1984, a total of 1,090,351,000 bushels of corn. For an example of quantity he explains that just one million bushels of corn that:

> loaded into railway cars, five hundred bushels in each, would require two thousand such cars, and make up a continuous train, eight miles long.

Murphy then attempts to have the reader imagine the enormity of 1.09 billion bushels which he says could be loaded forty bushels to a wagon allowing thirty feet for wagon, team, and space between wagons. The crop would stretch around the world six times. Or, if confined to the United States would "fill over forty-four continuous lines of wagon loads, from Boston to San Francisco!"[102]

Of this vast quantity of corn only four percent was exported in 1889. Murphy had tried to convince Europeans to learn to eat corn. His 1889 speech and his book of recipes in 1917 were attempts to encourage Europeans to incorporate corn as a staple in their diet.

As cross-fertilized corn was double-crossed to produce hybrids, the American farmer was pressured to produce larger and larger quantities of uniform corn which could be machine processed. The change from the small family homestead to commercial production began with hybridization of seeds and development of sophisticated machinery. The danger of planting only hybrid seeds was not recognized. In 1974, one major single variety of hybrid corn was planted across the United States and proved to be "susceptible to a fungus disease." The lost crop was "more than the consumption of China, Mexico and Latin America together."[103] If a similar problem were to happen with indigenous seeds the loss would be only 20 percent of the crop, while hybrid seeds could lose 90–100 percent of the crop.[104]

Enfield (1866) discusses the "luxury" of fresh corn, boiled or roasted, and fresh corn sliced from the cob for fritters, cakes, puddings, pies, and numerous preparations. The taste for fresh corn creates a "passion" for it which would develop in the food industry. The December 1862 issue of *Agriculturist* offered prizes for recipes on *"cooking* Indian corn meal," and 250 submissions were received.[105]

Emerson, writing in 1878, describes corn as roasted, parched, shelled and boiled whole, hulled, and made into succotash, hominy, samp, mush, and bread. Corn on the cob was reaching the city markets by this time. Green corn, so labelled because it was picked when the husks were still green, was eaten on the cob, or scraped from the cob and boiled with beans. "Americans commonly roast the ears before a clear fire, or on hot embers." Green corn was often scraped from the cob and dried in the sun or on the stove for winter usage, or scraped and salted. Succotash is corn scraped from the cob and boiled with beans. Parched corn was ground in a mortar and ground in a small corn mill, "it makes a very palatable meal, which, if one's beard is short enough, may be eaten dry; if otherwise, it is better wet, and best of all, wet with sweet milk." This parched corn is the same as used by travellers in pre- and post-contact times. Samp was made from cooking the coarser pieces of pounded corn. Emerson makes note that white corn was the best corn used for human food.[106]

Murphy's writings on 150 ways to prepare Indian corn are based mainly on yellow corn ground into corn meal. By 1917, American cooks had adopted corn into their daily diet and adapted their own familiar foods to create unique dishes. "Government War Bread 1917" was one-fourth wheat, rye, white, and corn meal with a tablespoon of salt, yeast, and water. Corn was cooked in milk, with the addition of cheese, vegetables, or meat, baked, boiled, and fried. What is called "Johnny cake" or "corn bread" was created with yellow corn meal, by adding baking powder, sugar, salt, eggs, and milk. Corn was an ingredient in "Virginia Corn Bread," and "Boston Brown Bread."[107]

Today Americans continue to eat corn as fresh corn on the cob, as canned sweet corn, corn bread, burritos, tortillas, corn chips, and, of course, corn flakes and many other corn cereals. The

Haudenosaunee continue to eat many dishes of parched, roasted, and hulled corn as well as the American versions of corn.

Corn-fed animals helped to build this country. As mentioned previously, 85 percent of corn grown was feed for animals which provided the American diet with corn-fed pork, beef, and chicken. These "efficient biological machines" converted corn into secondary products such as eggs, butter, cheese.[108] The farmers depended on the strength of work animals such as horses, mules, and oxen to pull farm equipment. Hardeman describes the "endless food chain" in which "horses, mules, and oxen produced corn; corn produced horses, mules, and oxen." When racehorses became popular they were called "cornfed dandies," because people believed that corn-fed horses were "superior to others in speed and endurance during hot weather or cold."[109]

In 1843, it was discovered that sugar, syrup, and molasses could be made from crushing the corn stalks in mills and boiling down the juice much like the maple syrup processing methods.

Making whiskey from corn had become an American custom. Emerson mentions whiskey as a medicine for asthma. Its value when distilled to alcohol helped preserve "substances from decay or decomposition." It was added to ink to prevent ink from freezing. In the Ohio army, the worst punishment for a soldier was to withhold his whiskey allowance. In 1794, Congress' attempt to tax whiskey was met with a rebellion of some 7,000 people. President Washington called out 15,000 men to quell the "insurrection." By 1850 there were thirty-five distilleries producing whiskey from corn, and by 1870 the number of distilleries increased to 141 and a booming business had developed.[110] Enfield viewed making whisky from corn as a gift "perverted to base and injurious uses." He makes an interesting observation about mankind's "twofold nature of good and evil," and expresses the expectation that man will not always be inclined to use only the beneficial purpose of a gift.

Between 1850 and 1870 the total number of manufacturers of starch from corn increased from 146 to 195. Hardeman says making corn starch was known to the early settlers "whose ancestors had made starch from other grains since Roman times." Cornstarch thickened soups, sauces, puddings, and gravy much as it is used today.[111]

Oil can be produced from the germ of the corn kernel. In 1866, Enfield recognized that 100 bushels of corn could produce sixteen gallons of oil.[112] By 1878, oil was a byproduct of the distillation process and four bushels of corn yielded two quarts of corn oil. Today corn oil is very much a leading oil used for cooking and salads, and boasts of being low in cholesterol.

Corn husks were used for paper making. Corn husks were useful as filling for mattresses. In earlier times corn husks were used as a packing material to pack fruit, for cigarette wrappers, and made into door mats. In Missouri a corn cob jelly was made by boiling corn cobs and making jelly from the juice![113]

By 1967, corn-refining processes enabled corn to be widely used for industrial products such as:

antibiotics	pencil erasers
aspirin pills	sandpaper
baking powder	dry-cell batteries
beer	paper
ice cream	books and bookbinding
ice-cream cones	crayons
cosmetics	chalk
soap	detergents
catsup	inks and dyes
chewing gum	explosives
licorice	oilcloth
marshmallows	paint and paint remover
peanut butter	photographic film
pickles	window shades
vinegar	matches
margarine	plastics
textiles	drinking straws

"One out of every three bushels of corn produced in America ends up in a refinery—three hundred railway cars full every day of the year."[114] The four basic corn products are corn oil, sugar, syrup, and starch, which are used in making the above products. It may even be possible that one day American automobiles will burn gasohol made from corn![115]

Much of the historical documentation recognizes that American Indians had many varieties and uses of corn, but suggests that they were superstitious, overly religious, and not as intelligent as settlers because they, supposedly, didn't have science.

Most of the improvement of corn was done by the American Indians before 1000 A.D., but the most *spectacular* changes have been made by the white man during the past two hundred years, especially during the past eighty years. The Indian did not understand corn with his head; he worshiped it in his heart, with a religious adoration. Undoubtedly, he took advantage of some *accidental* crosses which increased the size of the ears and the general quality of the plant. From a study of Indian languages of today, we may suppose that the Indians felt that their loving care and watchfulness, as well as their prayers to the corn god, had produced the changes for the better in corn. The Indian did not know the white man's science but, over the millennia, by patience and thankfully and prayerfully accepting favorable mutations, he changed the plant more radically than any other plant has ever been changed by man.[116] (emphasis added)

In a subtle juxtaposition, this paragraph both acknowledges the genetic engineering knowledge of Indians, yet determines that improved varieties were an "accident" tied to religious beliefs. The so-called superiority of "white man's science" produced "spectacular changes" in a mere eighty years. Yet, Wallace and Brown also lament the many varieties of corn that have been lost due to hybridization and they wonder what qualities these lost varieties might have contributed to the overall qualities of corn. The attitude expressed by Wallace and Brown is the standard view of Indians that generally takes up only a few pages in the beginning of books on corn.

In discussing the importance of corn to all aspects of American Indian life, Hardeman labels it the "deification of corn" and worship of "stone corn gods." Hardeman uses the term "corn culture" when referring to the pioneers and early farmers, yet when examining real corn cultures he reduces their beliefs to worship of stone gods, thus missing the opportunity for cross-cultural understanding.

But if American pioneers lacked some of the Indians' reverence for corn, they more than compensated by their worship of its bountiful yields and the material benefits from its ebb and flow in the channels of trade and commerce.[117]

There are two entirely different world views expressed in that statement. The first world view recognizes American Indian reverence for corn expressed in their understanding of the interrelationship of corn and human beings. The second world view, the pioneer's view of corn, is expressed in terms of productivity and monetary benefit, the underlying assumption being that only science and commerce are based on intelligent, reasoned thought.

Hardeman's nostalgic view of the loss of the early Americans' corn culture can be seen is this statement on the change from homesteading families to commercial production:

Late-twentieth-century Americans, when they are reminded of Indian corn, may conjure up fleeting mind pictures of shocks and golden pumpkins under fading sun and harvest moon, or steaming hot corn-on-the-cob dripping with butter, or perhaps girl and boy celebrating the finding of a red-grained ear with blush-faced kisses at a husking bee.... The reality of corn culture is now highly impersonal "factories in the fields," where tanklike tractors, gang plows, and combine harvesters cruise like armored divisions.[118]

Thus, we see a longing for the old days of small family homesteads just as we would find the same longing among the Haudenosaunee for a lifestyle previous to the devastation brought by modern society.

Summary

The role of corn in the interaction of cultures was a two-way process that affected aspects of many cultures. Corn as a central belief in cosmology and way of life was a widespread phenomenon in the western hemisphere. After 1492, corn was rapidly dispersed to Europe and Asia to become a commercial product that supported the slave trade and provided food for the poorer people around the world. Only in Africa was corn incorporated into an existing land-based culture and way of life.

Americans well into the twentieth century learned and practiced American Indian agriculture. Corn pervaded every aspect of agricultural development of this nation. American farmers began to change these Indian agricultural methods as plowing became a way to increase productivity. The face of American agriculture changed dramatically when uniformity of corn was introduced through hybridization, machinery was invented to plant and harvest crops, and when large-scale production replaced the family farm.

So important was corn to Haudenosaunee military strength, that two attacks almost a hundred years apart were designed to destroy the corn crops. The Quakers tried to change the corn culture of the Seneca by introducing plows, wheat, oats, rye, and the concept of private property. Being conservative, the Seneca selectively adopted pieces of the Quaker agriculture which would not upset their cultural world view and way of life. Evidence of this selective adaption is apparent when Sullivan's armies found fruit orchards with 1,500 trees in 1779. The agreement to try an experimental plot with both traditional hills and plowed rows illustrates Seneca conservatism, along with willingness to experiment.

Cookbooks during World War I abound with recipes, and how-to instructions on cooking Indian corn. A patriotic appeal encouraged people to grow vast crops and eat corn to support this Nation during World War I. Corn was used to feed domesticated animals and the products from these animals increased the wealth of the American farmer. Corn was also used for making whisky, corn starch, corn oil, corn sugar and syrup, and other corn products.

The Haudenosaunee and other indigenous peoples' use of corn was a significant aspect of their world view and was interdependent with their cultural way of life, but American farmers' utilization of corn was intended for commercial purposes. Today we see the disappearance of the small American farmer who has been replaced by gigantic corporate entities. The Haudenosaunee agriculturalist, however, can continue small-scale agriculture much the same as their ancestors. Indeed, the same seed survives today.

Chapter 8

Dynamic Aspects of Haudenosaunee Culture

All cultures change, none is static, and curriculum materials can present the dynamic nature of change by showing how the culture incorporates or adapts items from the modern world but maintains the continuity and integrity of its world view. Ray Gonyea (Onondaga) describes this process as "selective adoption" in which fabrics, metals, and new materials are "adapted into a cultural framework and the end product was distinctly Indian."[1]

Some missionaries and state officials saw this as extreme conservatism or resistance to change and "condemned the traditional governmental leadership."[2] The Haudenosaunee people are extremely selective about adopting items from the modern world, which is a major reason why they continue to exist as an identifiable people, that is, Haudenosaunee conservatism has meant their cultural continuance. They not only know their ancient world view but continue to live by its tenets within the modern context.

The first section examines the dynamic nature of Haudenosaunee culture as portrayed by Seneca artist Ernest Smith. Between 1935–41, Ernest Smith painted more than two hundred views of Haudenosaunee culture while employed by a Works Progress Administration (WPA) project through the Rochester Museum. In this section we examine the major influences on Ernest Smith to assess whether his paintings accurately portray Haudenosaunee beliefs and way of life.

The next section places the Ernest Smith paintings into the sequence on Haudenosaunee world view: Thanksgiving Address, Creation, Great Law of Peace, Handsome Lake, and the thematic focus on corn. The paintings are analyzed to identify evidence of pre-contact, post-contact/selective adoption, and continuity that supports the dynamic nature of Haudenosaunee culture.

The final section examines the need to show the dynamic nature of cultural change and continuity into contemporary times, which requires photographs from 1935 to the present, and contemporary works of art.

Influences on Ernest Smith (1907–75)

Ernest Smith was born in a log cabin on the Tonawanda Seneca Reservation in 1907, the youngest of seven children. "He quit grade school before finishing in order to support an aging mother."[3] His Seneca name, *Geo-yaih,* means "From the Middle of the Sky." At the age of twenty-eight he went to work for Arthur Parker, director of the WPA project at the Rochester Museum, Rochester, New York. Through this project, which employed about 100 Seneca artists and crafts people from Tonawanda and Cattaraugus Reservations, approximately 5,000 works of art were produced between 1935–41. In 1972, Ernest Smith was recognized by his own people at the Iroquois Conference where he was given the first "Iroquois of the Year" award "for contributions to the preservation of the Iroquois way of life through the medium of his art."[4]

Ernest Smith had available to him on a daily basis the knowledge of many Haudenosaunee Elders. A photograph of the Seneca people participating in this WPA project from the Tonawanda Seneca reservation shows Ernest Smith to be much younger than most of the other artisans.[5] One of those elders was Jesse Cornplanter (1889–1957), who was born and raised in the longhouse tradition at the Newton longhouse at Cattaraugus.[6] Arthur Parker's *Book of Maize* (1910) contains illustrations Jesse Cornplanter sketched while still a youth, indicating that the two had a well-established working relationship prior to the WPA project. Indeed, some of Ernest Smith's oil paintings are beautiful renditions of the youthful sketches of Jesse Cornplanter.

Another major influence on Ernest Smith was Arthur C. Parker (1881–1955), a Seneca from Cattaraugus, who designed the WPA project and served as director of the Rochester Museum. Lewis Henry Morgan collected a vast amount of Iroquois materials during the 1840s, through his Seneca informant Ely S. Parker, who was Arthur Parker's great-uncle. This ancient collection of material objects of Haudenosaunee culture was destroyed in a disastrous fire at the Albany Museum, which Arthur Parker, an archeologist at the New York State Museum in 1911, witnessed.

During his career, Arthur Parker was an "archeologist, ethnologist, folklorist, historian, author of children's books, defender of Indian rights, and museum administrator."[7] He came from a quite well-educated Seneca family. His mother and grandmother were not Seneca, therefore he was not recognized in the traditional matrilineal way as a Seneca, but when he was older the Seneca adopted him into the Bear Clan giving him the Seneca name *Gawasowaneh* or "Big Snowsnake."[8] Arthur Parker's childhood home was a natural gathering place for Seneca elders and Parker grew up listening to the Seneca oral traditions as related by his grandfather, Nicholson Parker, and his great-uncle, Ely Parker.

Ely Parker, *Ho-so-no-an-da* (1828–95), was the son of William and Elizabeth Parker. William was a young child when his family relocated from Allegany with Handsome Lake to Tonawanda. His uncle, Samuel Parker, was one of the fourteen Seneca chiefs. It is said his mother Elizabeth was a descendant of *Jigonsaseh*, the Peace Queen, who was the first woman to accept the Great Law of Peace. The boys grew up hearing the oral traditions, and living the Haudenosaunee way of life, that is, they were raised in the old Indian ways. During his lifetime, Ely saw bark houses replaced by log cabins, and buckskin clothing replaced by cloth. When Ely was ten he moved to Grand River reservation in Canada to get away from the land disputes caused by the fraudulent 1838 Buffalo Creek treaty. He stayed there for two years, learning the ceremonial ways of the longhouse. He returned to Tonawanda determined to master the English language and defend his people. His skill with the English language enabled him to serve as an interpreter and by the age of fifteen he was often sent to Washington with the Seneca chiefs to translate negotiations in the effort to retrieve the Tonawanda land lost in the Buffalo Creek treaty.[9]

Ely met Lewis Henry Morgan and served as his informant, collaborator, and interpreter while Morgan conducted his field research for his 1851 book *League of the Ho-de-no-sau-nee, or Iroquois*. Morgan was only thirty when the *League* was published and Ely was still a youth. Morgan's publication of the *League* gained him the reputation as the "father of American ethnology."[10] Ely assisted Morgan by translating the Elder's knowledge from Seneca into English and by gathering the vast collection of

Haudenosaunee material culture that Morgan placed in the New York State Museum.

Ely went to law school for three years only to find that he could not be admitted to the bar because he was neither a "male white man" nor a citizen.[11] He became a civil engineer and worked on the Erie Canal. Ely Parker served in the Union Army as the military secretary to General Ulysses S. Grant, and helped write the terms of surrender that ended the Civil War. He was the first Commissioner of Indian Affairs, and worked for the New York City Police Department.[12] Researchers have generally focused on his accomplishments in the white world, while this study is focused on his life in the Seneca world, because his knowledge of Haudenosaunee culture was extensive and he witnessed great changes in his lifetime.

Ely Parker's influence on Arthur Parker was significant and inspirational. Arthur's influence on Jesse Cornplanter and Ernest Smith was significant during the WPA project. Thus, a documented direct line of cultural knowledge and scholarly work can be traced for a hundred years prior to the Ernest Smith paintings.

Arthur Parker wanted Haudenosaunee culture to be presented with respect and dignity. He was a cultural relativist, and even though he turned down the opportunity to study with Franz Boas, he understood Boasian principles, perhaps because he had such a deep respect for his own culture, and he viewed Haudenosaunee culture from a Haudenosaunee perspective. His ethnological work presents the Seneca way of life with dignity even though his great-uncle Ely had collaborated with Morgan, who had been instrumental in establishing the social evolutionary scale to describe cultures. Perhaps Arthur Parker knew that Morgan misinterpreted Haudenosaunee culture because Morgan was unable to step out of his ethnocentrism. Arthur Parker's writings do, however, reveal a thread of the social evolutionary scale in some of his vocabulary. In that sense, Arthur Parker might be viewed as a transitional person between the academic philosophies of Morgan and Boas.

Jesse Cornplanter and the many Elders on the WPA project served as cultural resources for the young artist Ernest Smith. He was surrounded by vast stores of cultural knowledge from which to check his perceptions while illustrating aspects of Haudenosaunee culture. His paintings represent the knowledge and wisdom of the ancient Haudenosaunee culture handed down by the Elders, generation to generation, through oral traditions.

Why the Ernest Smith Paintings?

The Ernest Smith paintings transmit cultural knowledge that has significant meanings in the Haudenosaunee world view. The paintings cover a vast time period from the antiquity of the Creation Story, to the historic Sullivan campaign, to the continuity of traditional ceremonies in the 1930s. Ernest Smith provided visual images of the important elements of Haudenosaunee culture—the ancient beliefs, stories, elements of lifestyle, and historical events. The paintings contain evidence of cultural adaptations, or selective adoption of elements from the non-Haudenosaunee world, which are incorporated into the lifestyle without harming the original world view. The paintings can be extremely useful in the classroom setting to provide visual imagery of Haudenosaunee world view, lifestyle, cultural adaptation, continuity of world view, and interaction of cultures.

The paintings are important to dispel stereotypes of savage Indians, generic Indians, and Indians as fossils. However, one limitation of the paintings is the lack of a visual image of daily life in 1935. Photography from the time period can be utilized to provide those visual images.

This writer is not an art historian and as such was not equipped to analyze the paintings on an artistic level, therefore only a few comments are offered on the artistic style. Ernest Smith used oil, watercolor, and tempura mediums. He tried varying styles including landscape backgrounds, and a style with a central figure as the focus with little or no background. There are four paintings identified as *Fantasia*. In trying to understand why Ernest Smith would experiment with the *Fantasia* style to illustrate Haudenosaunee stories, I rented the video of Walt Disney's *Fantasia*, which we viewed as a family. As we watched, my children, who have been raised in the ways of the Longhouse, said to me, "Oh, look Mom, it's the Little People." Thus, I observed their comments, thinking that their reactions would be similar to Ernest Smith's, and attempted to see what he saw in *Fantasia*, in terms of world view. In *Fantasia* the little elves and fairies make nature beautiful, which is similar to the Little People's responsibility in Haudenosaunee culture. It would have been interesting to talk with Ernest Smith about this cross-cultural exchange and why he experimented with the *Fantasia* style. However, artistic style is certainly an intriguing area which hopefully a future scholar will examine.

Each painting was examined to determine the following elements:

1. What insights into Haudenosaunee world view and epic narratives do the paintings provide?

2. What time period is represented?

The pre-contact era can be identified by buckskin clothing, pottery, birchbark trays, and bark longhouses, and the post-contact era by the use of cloth clothing, iron kettles, log cabins, plank board houses, and metal tools. Or is the evidence of time periods intermixed, i.e., cloth clothing and pottery? How clearly does he distinguish between pre-contact and post-contact? Is this significant?

3. What types of groupings of people are provided? Do these groupings illustrate the communal nature of the Haudeno-saunee lifestyle? Are families (mother, father and children, and extended family) and community gatherings portrayed or is the emphasis on individuals?

4. What symbols of adaptation and continuity are represented? Evidence of continuity and selective adoption can be found in the use of cloth and beads to make ceremonial clothing that formerly would have been made from buckskin with porcu-pine quill decorations. Selective adoption is a critical issue in examining the paintings. Is there evidence that adapting an aspect of the white world changed the world view? Or is there evidence of selectively adopting an article from the white world and adapting it to fit into the Haudenosaunee world view?

The paintings were placed into categories that follow the world view presented in the case study (see Chapters 5 and 6).

Thanksgiving Address	The Code of Handsome Lake
Corn as a Cultural Focus	Ceremonial Cycle
The Creation Story	Historical Interaction
The Great Law of Peace	Cultural Stories

Within each category, the paintings illustrate and interweave the fundamental Haudenosaunee themes of thankfulness or appre-ciation, and an understanding of the reciprocal relationship be-tween these elements and human beings. Many of the paintings are stories that recall oral traditions and express values. To teach with these paintings, the teacher must have a background in

Haudenosaunee symbology and the interrelated nature of many aspects of the culture from a Haudenosaunee perspective. Ernest Smith provides the visual images of that interconnectedness. (See appendix for a complete list of the paintings.)

The Thanksgiving Address

Each section of the Thanksgiving Address, except for the Sky World spiritual entities the Four Beings and the Creator, was illustrated by Ernest Smith. Due to the interrelated nature of the Haudeno-saunee world view a single painting includes many aspects of the culture, therefore the painting may be referred to in several sections. Many of the paintings illustrate a story that I refer to although space limitations prevent telling the entire story. Some of the paintings illustrate a significant event which is briefly described. The number following the title indicates the number on the painting in the Rochester Museum.

The People

Paintings in this section illustrate aspects of Haudenosaunee lifestyle and the attitude of respect and thankfulness. Smith shows the connection to nature in "The Child Studying Nature" (1102), in which a young boy sits quietly observing a raccoon fishing, as two deer cross the stream in the background. "Children at Play" (1116) shows two young children, with cloth clothing, who have constructed a small village with a stockade and moat around it, and they play with corn husk dolls. In the painting "Lunchtime" (1108), the mother or grandmother is calling two children, who appear to be racing toward her, to eat corn bread and corn soup from pottery containers, and they wear buckskin clothing. A sad painting, "Adoption of Orphans" (1119), shows a family leaving the site of two graves, probably the children's parents. The little girl is weeping. The adults point to two children running toward them from the village, as if to say, "here's your new family."

> There is always a family to take care of an orphaned child. *The moral here is there is unity among the people.*
>
> *Ernest Smith*[14]

A very tender scene, "Mother and Baby" (736), shows a mother in traditional-style clothing made of cloth, rocking and singing to her baby who is asleep on an elaborately decorated cradleboard. "Clean Camps" (1112) shows Haudenosaunee women in cloth clothing sweeping out the longhouse and dumping scraps into a refuse pit in a very neat, clean, and orderly village. In "Storyteller" (1050), an older man, in buckskin clothing, sits under a tree talking to a teenage boy. The paintings in this section illustrate the connection to nature, family values, care of orphans, cleanliness of the village, feeding the children and listening to Elders.

Mother Earth

Haudenosaunee understanding of the reciprocal relationship of the earth, animals, and humans can be found throughout all of the paintings, therefore, only a few significant paintings are cited here. In the painting "Conservation" (75.224.1), a father and son in buckskin clothing are walking away from a stream, each having caught two fish. The boy looks back at the stream as if to say there's more fish, but the father puts up two fingers as if saying we have what we need. In "The Potter" (685), a woman in buckskin uses clay from the earth to make pottery as a man tends the fire where the pots are being fired. "Two Men Dipping Bowls" (691) is a sequel to "The Potter" that shows the men removing the pottery from the fire with sticks, placing the pots into a bark tray full of water to cool. The above mentioned "Child Studying Nature" shows the boy, raccoon, water, fish, and deer as balanced elements of the same ecosystem.

Bodies of Water

Human beings' use of water for drinking is depicted in several paintings: a man dips water with the flow of the stream (704), and two smiling women wearing cloth clothing carry water in pottery vessels (741). The Mother Earth section cited paintings that incorporate water for making pottery. The use of bodies of water for fishing are depicted in "The Fisherman" (1131), which shows two men in buckskin fishing, and the previously discussed painting "Conservation."

Grasses/Plants

The grasses and plants are utilized by the people for food and medicine. "Making Sunflower Oil" (1022) portrays two women in buckskin clothing, one using a corn pounder to pound the sunflower seeds, while the second woman is using a grinding stone to grind the seeds. In the background is a large pottery container nestled on stones with a fire under it to boil the oil from the pounded seeds.

Splint baskets are used in "Gathering Artichokes" (1002). "Gathering Oyster Mushrooms" (1003) shows two women in buckskin clothing picking mushrooms from an old log. In the background are drying racks and baskets, which illustrates food preservation. "Medicine Picking" (719) depicts a man in cloth clothing and a woman in buckskin with carrying baskets. They are burning tobacco, which is done to offer a thanksgiving to the Creator for the medicines they are about to pick. The paintings in this section illustrate the use of natural foods gathered from the environment and their preservation for food and medicine.

Hanging Fruit

Strawberries are an integral component of the Haudenosaunee ceremonial cycle and a thanksgiving is offered each year at the Strawberry Ceremony when the wild strawberries appear, because there exists the belief that one day the world will be so polluted that the wild strawberries will no longer grow. Also, in death, Handsome Lake said the road to the Creator's land is lined with wild strawberries. "Jungie Turning Strawberries" (1688) represents the Little People whose job is to turn the fruit so it ripens evenly. A happy, smiling family in buckskin clothing is shown using birchbark baskets in "Picking Wild Strawberries" (1010). A mother and daughter wear ceremonial cloth women's dress, and use splint baskets for containers in "Picking Black Raspberries" (1019).

Trees

The trees provide warmth, food, and shelter. The head tree is the maple. Ernest Smith provides four paintings on the theme of the maple tree in Haudenosaunee world view. "The Spirit of the Wind Meeting the Spirit of the Maple" (653) depicts the wind as a male who swoops down from around a tree and the maple tree as a female, in buckskin clothing, who seems to step out of the tree trunk. This illustrates a basic belief in Haudenosaunee world view, that all elements of nature are alive, they have a spirit. "Giving Thanks to the Great Spirit for the Maple Sugar" (716) shows the people of the village gathered for the Maple Syrup ceremony which is held outside a bark longhouse. The people are dressed in cloth clothing and a pottery vessel of maple syrup simmers over an open fire. The combination of a bark longhouse with cloth clothing may have been illustrated this way to indicate the antiquity of the Maple Syrup ceremony, i.e., bark longhouse, and the contemporary continuity of this ceremony, i.e., cloth clothing. "Maple Sugar Time" (709) shows a man and woman in buckskin clothing. The man is gathering the sap which has dripped down wooden spouts into pottery and wooden bowls. The woman is kneeling next to the fire over which maple sap boils in three large pottery containers. Interestingly, Smith paints four large flat rocks behind the fire which seem to reflect the light and warmth of the fire in a background of woods with snow still on the ground. "Maple Sugar Implements" (1133) provides sketches of the tools used for making maple sugar.

Other spiritual connections to trees are illustrated, which provide the sense that all elements of nature are alive and have a spirit.

> The Frost Spirit is revealed when he strikes the trees with his war club. He is called Hah-t'ho, and his body is formed from clear ice. His breath forms the snapping winter mist when the temperature is below zero (655).[14]

"The Spirit of the Pine Tree" (680) illustrates the spirit as the face of a man in the top of the pine tree. An older, gray-haired woman gazes upward at him. Illustrating spirits as the intertwining of nature and human characteristics was one of Ernest Smith's strengths in conceptualizing the connections between humans and nature.

Two women in ceremonial clothing use splint baskets to gather plums and spread them out on a mat in "Drying Plums" (1012). In late summer or early fall plums are picked and dried to be later added to other foods, or boiling water is added to make a drink. "Making Hemlock Tea" (1130) and "Making Hickory Nut Oil" (1021) illustrate additional utilization of trees. "Gathering Chestnuts" (1020) includes a mother, father, young boy and his dog. They wear buckskin clothing, and use splint baskets to gather the chestnuts. The boy is using a grinding stone to make the chestnuts into flour. The dog is looking up into the tree as the father hurls a large piece of wood up into the tree to knock the chestnuts down.

"Longhouse Under Construction" (683) shows three men, wearing buckskin leggings and breechcloth, building a peaked roof bark longhouse. Splints were pounded from black ash logs to make splints for "The Basketmaker" (711).

The many uses of the tree for food, warmth, baskets, foods, and shelter, and the appreciation expressed in the Maple Syrup ceremony provide insightful images of Haudenosaunee world view.

Animals

The paintings in this section provide several perspectives on the interrelationship of humans and animals. First animals and human beings are related, and second, animals give their lives to humans as food. "The Story of Red Hand" (743) shows a young Haudenosaunee man fixing a splint for a small tree while all the animals watch him. A reciprocal relationship is shown in the story about Red Hand and his respectful mind towards animals, when Red Hand is wounded and the animals make a medicine to heal him, as illustrated in "The Little Water Society" (715). "The Hurt Warrior" (955) shows a bear tenderly holding the arm of an injured warrior as an eagle watches over him.

> The Chipmunk is pictured eating a black walnut from a bowl of corn soup. The ladle is carved with a bird representing the Hawk Clan on the mother's side. Mr. Smith conveys the idea of friendship and sharing with all, even the animals of the forest. (737)[15]

"Thanking the Spirit of the Bear" (1131) and "The Fisherman" (1078) show an understanding of the reciprocal relationship of humans and animals. "The Bear Dance" (732) shows a ceremonial dance in which the bear is represented by a man wearing a bear hide tied tightly around him. He leans over to eat berries from a basket on a bench in the longhouse. The dancers, men and women dressed in cloth clothing, are accompanied by one singer. This painting provides elements of the ancient ceremony.

The second aspect, the use of animals for food within an interdependent Haudenosaunee world view is illustrated in "The Cycle of the Hunter and the Hunted" (1066).

> The balance in nature is maintained through a constant cycle of the hunter and hunted from the lower forms of animal life up to man. Nature's success depends on the proper balance of this cycle.[16]

The cycle begins with a rabbit running away from a fox, a bobcat in a tree is poised to leap on the fox, and a hunter has his bow aimed at the bobcat. "The Hunter" (1049) shows a hunter and a deer, and "Preparing Smoked Venison" (1001) illustrates the use of drying racks. "The Story of the Hunters" (950) shows a father and son in buckskin clothing hunting wild turkeys. "Fighting a Bear"

(1052) shows a man in buckskin clothing with a stone hatchet fighting a bear.

Other connections to animals are shown in the vast number of stories in which humans and animals interact, and sometimes humans even transform into animals. "The Wounded Bear" (1088) tells the story of the three hunters who chase the bear into the sky where it becomes the Big Dipper.

Birds

"How the Birds Got their Color" (714) shows a bird sitting on a wolf's nose with many birds in the trees. "The Vulture Chooses His Feathers" (1064) is described by Smith:

> He (the vulture) had the choice of all the different kinds of feathers, but he was in such a hurry to fit himself with nice plumage that he put the outfit on and it's too short for him. . . . He was in such a hurry to pick out a cloak that he picked the wrong one. . . . I think the moral is, "don't be too fast to pick out something that may not suit you."[17]

There are three paintings that tell stories of the eagle: "Dance of the Eagle" (619), "Eagle Dance" (494) and (727). "The Pigeons" (1107) shows a man chopping down a tree full of pigeons. At the time of contact, passenger pigeons were so plentiful that they would darken the sky. Today they are extinct. In the "Nest Robber" (1129), a mother is severely scolding a young boy who is caught while up in a tree stealing robin eggs. "Hunter Waving to Geese" (75.225.1) shows respect and friendship.

The Three Sisters—Corn, Beans, and Squash

Known as *Diohe'ko*, "our sustenance," or "they who support us," the "Three Sisters" (721) are presented in the painting as female spirits. The corn spirit stands within the corn which appears to originate from her feet. The skeletons of fish, as fertilizer, can be seen just below ground level. The bean sister has her arm around the shoulder of the corn sister. The squash sister sits on the ground behind the squash plants. At the very edge of the painting is the Little Person whose job is to turn the blue hubbard squash so it ripens evenly. The same painting is repeated but omits the Little Person (1014).

The Thunderers, Our Grandfather

"He'no, The Thunder God" (724) appears as a spirit in the clouds above a lightning storm. He is poised to shoot a bow and arrow. His mouth is open wide as if he is shouting.

> While traveling, Good Mind (the twin) met the Thunder God who promised to help mankind by searching for and destroying evil spirits. In the Spring he brings the rain to nourish the seeds.[18]

In Parker's description of the thunderers, "His flint-tipped arrows made the lightning . . . his voice was the thunder's roar."[19]

"The Hail Spirit" (633) throws hail from a pottery container as a man covers his ears. In "Thunder God and the Spirit" (656), the Thunder God uses lightening to destroy a horned serpent.

After a storm, the rainbow appears. "Searching for the Rainbow" (1077) shows two men in buckskin clothing running toward the rainbow. The story relates that He'no (thunder) caught the serpent and stretched him across the sky to show that he had defeated the serpent once again.

The Elder Brother Sun

In the Creation Story, the Good Twin asked his Grandmother where to find his father and she told him to travel to the east. Ernest Smith illustrates the epic battles the Good Twin endured to reach his father, the Sun. "Braving the Sea" (1067) required defeating a water serpent. "Swimming the Cataract" (1068) shows him dealing with swimming up a huge waterfall. "Removing the Boulders" (1069) shows him throwing large boulders out of the way. "Going Through Fire" (1070) and "Blown by the Wind" (1071) were the last two adventures on this journey. "The Meeting" occurs between the Good Twin and the Sun Father (1072). Utilizing the sun's energy to dry foods is presented in "Venison Drying" (1011) and "Plums Drying "(1012).

Grandmother Moon/Stars

The Grandmother Moon is responsible for waters such as dew, and the tides, and she regulates the woman's monthly cycle and the birth of children. "Woman in Seclusion" (1109) shows a village, with a wigwam-type structure separate from the village and one woman sits by the wigwam. This represents the time when women would be separated from the village during their monthly cycle. The importance of children in the Haudenosaunee world view was discussed in the first section on People.

"Wounded Bear" (1088) explains the origin of a constellation that becomes the Big Dipper, and "The Seven Dancing Stars" (1015) tells the story of seven young boys whose parents refused them food. They sang a song and rose into the sky to become the Pleiades.[20] The "Elk That Fell in Love with the Morning Star" (651) shows an elk and beautiful woman in buckskin clothing, but a story has not been located for this painting.

The Four Beings

Ernest Smith did not illustrate the Four Beings, who are spiritual, or Sky World entities.

The Wind

The Wind appears in two contexts. "The Spirit of the Wind (male) Meeting the Spirit of the Maple (female)" (653) was discussed in the tree section. "The Daughter of Sky Woman" (742) illustrates the daughter who was impregnated by the West Wind and gave birth to the twins.

Handsome Lake

"Handsome Lake and the Three Messengers" (630/631) illustrates the story of the messengers who brought Handsome Lake a vision about survival of the old beliefs in the modern world context.

The Creator

Ernest Smith did not attempt, nor has any Haudenosaunee artist attempted, to illustrate the Creator.

The Ernest Smith paintings illustrate the Thanksgiving Address, Creation, and ceremonies, which are integral elements of Haudenosaunee world view. There are paintings that are pre-contact, that is, the people wear buckskin clothing, and use splint baskets, bark bowls, and the ancient pottery. The diet of the Haudenosaunee people consisted of foods from fishing, hunting, gathering, and agriculture which are easily identified throughout the paintings. Children were highly valued as illustrated in many paintings. Their education was obtained by participating in daily activities with adults.

There are paintings that show post-contact period by including cloth clothing and bark longhouses. "Giving Thanks to the Great Spirit for the Maple Sugar" (716) show the ancient ceremony with people dressed in cloth clothing outside a bark arched roof longhouse. These depictions of ceremonies provide information on the antiquity and continuity of the ceremony by showing the bark longhouse, and selective adoption of cloth for making clothing.

Family groupings and happy smiling people are evident in several paintings. The village community attends ceremonies. Individuals are more likely to be portrayed in the hunting scenes. Thus, we find evidence of the extended family and community culture with only a few individual type activities.

The paintings show the understanding of the interrelationship of human beings, animals, and nature as given by the Creator. Their coexistence is essential to Haudenosaunee world view.

Corn as a Cultural Focus

Ernest Smith illustrations depict corn in many aspects of Haudenosaunee life including the Creation, ceremonial cycle, Handsome Lake, preparing corn, and stories.

In the Creation story, as Sky Woman is falling from the Sky World, streaking toward her through the dark sky is a panther-like being with a long white streak behind him like a comet (545). This creature, Gaha'ciendie'tha' or Ga:syoje:ta' (in Seneca) has been interpreted as a blue-black panther creature with white light streaking out behind him that comes habitually. This creature is from the Sky World. He has caused the jealousy in the Chief who pushed his wife, Sky Woman, through the hole in the Sky World caused by the uprooting of the celestial tree. As she's falling Gah'ciendie'tha' gives her an ear of corn, a mortar and pestle, and a small pot. He says to her, "This, verily, is what thou wilt continue to eat."[21] We know corn was evident in the Sky World from the Creation story discussed in chapter 5.

The two paintings of the spirits of the Three Sisters were discussed in the Thanksgiving Address section. In the planting ceremony they are appreciated in several ways. Since agriculture is the women's responsibility, they have a special Bread Dance or "Women's Dance" (610) in which the lead woman carries a small turtle rattle, and the next woman carries braided corn, or seeds of corn, beans, and squash, and each woman sings her song as they dance to express gratitude for the corn. There is a large iron kettle of steaming hot corn soup around which the line of women circle.

Smith illustrated the women wearing the ceremonial cloth dresses, which are the same style worn today for ceremonies.

During this planting ceremony new seed is exchanged among the planters to keep the seed strong. "Introduction of New Corn" (1084) shows a man with a bundle of corn on his back, handing ears of corn to a family that includes the parents, daughter, and grandmother. "Blessing the Cornfield" (511) shows a very ancient custom when a naked virgin would "scatter a few grains of corn to the earth as she invokes the assistance of the spirit of the corn for the harvest."[22] The antiquity of this custom is evident by the bark longhouse, splint baskets, and pottery. Before the MidWinter ceremony begins the Bigheads go to the homes to announce the beginning of the ceremony. " New Year's Announcement" (497/712) show the two men wearing buffalo robes, tied at the waist and ankles with braided corn husks, and carrying a pestle. They are approaching a log cabin and in the snow two tracks appear, perhaps from wagon wheels, thus this painting was post-contact period and illustrates the continuity of the announcing of MidWinter ceremonies. "The Dance of the Husk Mask" (712) takes place during MidWinter ceremonies. The dancers wear masks made of corn husks and carry a braid of corn. "The Eagle Dance" (494) shows dancers in both cloth and buckskin and there is a corn pounder in the corner of the longhouse.

Handsome Lake tells of walking through a cornfield and feeling the Corn Spirit touching him and asking to go with him, but he admonishes her to stay with the people. The "Corn Spirit" (690) illustrates this scene, which was also illustrated earlier by Jesse Cornplanter.[23]

"Hoeing Corn" shows three women in very nicely beaded cloth outfits hoeing corn with hoes made of a blade of bone attached to a stick. This painting combines the beautiful clothing, which certainly wouldn't be worn while hoeing, with the ancient type of hoe. Two women return from the field with corn bundles on their backs in "Corn Gatherers" (744). "The Story of Red Ear" (516) tells of good luck for the person who found the red ear. In the painting, two women appear to be laughing and having a good time as they braid corn. "Roasting Corn" (1104/1114) presents the method of preserving corn by husking it and leaning it on a rock near the fire to brown. The roasted corn is shelled and placed in three bark baskets to finish drying. "Working Corn" (1100) by scraping it off the cob with a deer jaw bone and placing it in a bark basket to dry is one method of preserving green corn. The traditional method of removing the hard outer hull from the corn by boiling it with wood ashes and then pouring it into a splint corn basket to wash away the ashes and hulls is illustrated in "Washing Corn" (740).

A very funny story of "The Boy Who Popped Too Much Corn" (1023) shows the young boy running away from a bark longhouse that is exploding at the seams with popped corn! A more serious painting, "Denial at the Storehouse" (1127), shows the results of not having assisted in the communal work. The man was refused corn, although that was only done as a extreme measure to teach a lesson. A Tuscarora corn story told of throwing of corn into the fire and the woman, who is the Corn Spirit, leaves the people. "Corn Maiden is Burned When Corn is Burned" (1685) illustrates this lesson, which teaches the people to appreciate the corn. "Praying for Manna" (1065) shows three people with baskets sitting outside the bark longhouse. Corn appears to be falling from the sky. This could be referring to the Sky World when the chief rains corn into Sky Woman's village as part of the marriage arrangements.

The Creation Story

Ernest Smith provided this description of the Haudenosaunee Creation story in his painting "Sky Woman" (545):

> This is the sky world on top of the picture. It shows that through jealousy he (the Great Chief) uprooted the tree making a hole in the clouds and pushed his wife through the hole. . . . The flying panther is the symbol of the flying star (a sign of disaster). . . . This just shows that it was night. In the sky world it is always sunshine. The geese caught the woman and let her down easily on the big turtle's back. There was no earth before that time. It was just water. The Indians used to tell that the earth is actually held up by the great turtle.[24]

In his second painting of "Sky Woman" (1011), Smith shows just one bark longhouse in the Sky World and omits the flying panther. The "Daughter of Sky Woman" (742) is impregnated by West Wind which is shown as a spirit. "The Twins" (1092) born to Sky Woman's daughter are illustrated and there is beam of sunlight on one of the boys. The "Great Feather Dance" (618) illustrates the Creator's dance, which is a vital part of each ceremony. In Smith's painting the dance is taking place inside a wooden building, with benches, and looks very much like present-day longhouses. The people are dressed in cloth clothing as they dance around the singers with their turtle rattles. Another gift from the Creator was the game of lacrosse, which was part of healing medicine. Smith illustrates another use of the "Lacrosse Game" (696) as sport played at large intertribal gatherings.

The Great Law of Peace

"Hiawatha's Sorrow" (613) shows Hiawatha dressed in buckskin as he grieves at his daughter's grave. In the background are the graves of six other daughters. The "Peace Queen" (546) presents Jigonsaseh at a longhouse where it appears two groups of people are conducting a meeting. The leader of one group wears a panther skin headdress, which indicates the Erie Nation. The men in the second group have Mohawk style haircuts. All are dressed in buckskin and one man's breechcloth is decorated with corn. The two groups meet with Jigonsaseh in the middle, and one man near her drops his war club. The "Council with Tadodaho" (616) shows Tadodaho sitting on the ground with snakes in his hair and a war club in his hand. The Peacemaker and Hiawatha are speaking with him. Wampum belts, strings of wampum, and a flint tipped arrow lay on the ground. All are dressed in breechcloth and leggings. Another interesting visual image is presented in "Totadaho, Hiawatha, and the Peacemaker" (723), because the Hiawatha wampum belt which represents the union of the Five Nations into the Confederacy has a heart in the middle instead of the standard pine tree. The origin of wampum took place during the Great Law of Peace epic and "The Wampum Maker" (681) represents that origin as he weaves the Hiawatha Belt.

The Code of Handsome Lake

Handsome Lake and the Three Messengers were described in the Thanksgiving Address section. "The Vision" (524) presents a spirit-like village of longhouses in the trees, in which people are moving about. Two men sitting under a tree are observing this spirit-like village. Handsome Lake is shown inside a plank board longhouse in "Handsome Lake Preaching" (707). He holds wampum strings, and the men are seated on the benches behind him as he stands to talk, and the women are seated on the opposite side. The sun streams through the three windows. Everyone wears cloth clothing.

The next group of paintings apply to sections of the Code of Handsome Lake. "The Gluttons" (958) shows three fat men and one starving man. "Spirit of the Gambler" (733) shows a blanket, bench, bowl game, and the spirit of a man looking back at these items. "Pathway to the Souls" (1013) and "The Trail to the Great Beyond" (1082) represent the paths to the Sky World after death.

Ceremonial Cycle

The paintings referring to ceremonies and ceremonial events, such as Maple Syrup, Strawberry Dance, Planting Ceremony, Maple Syrup Ceremony, Great Feather Dance, BigHeads announcing MidWinter, Women's Dance, Husk Dancers, and Handsome Lake have been discussed previously in the appropriate sections presenting those aspects of the world view.

There are many more paintings that illustrate additional ceremonies intended to heal the sick. These ceremonies are generally conducted by specific people who belong to a specific medicine society or group responsible for knowing the ceremony, songs, dances, and requirements. Unless a person has need of this ceremony or belongs to a specific society, that person usually doesn't know very much about that society. Only those who need these services are involved. Therefore, the paintings pertaining to these ceremonies have been included in the index but will not be discussed, to respect the privacy of these spiritual and medicinal ceremonies.

Historical Interaction

Smith illustrated the importance of trade, including both pre-contact Indian-to-Indian trading, and post-contact trade with the Dutch. There are four paintings of the Sullivan campaign, and an intriguing painting called Progress.

Pre-contact trade systems were well established. "Trading Arrowheads" (1110) and "The Stockade" (1114) depict trading for furs with other Indian Nations. The "Gift of the Panther Skin" (954) represents trade for furs with the Eries, who were known as panther people. "The Trappers" (1128) shows one Indian on the shore with a pack of furs on his back and another Indian approaching in a canoe.

Post-contact trade is evidenced in "The Dutch Trader" (1117). The Dutch man wears a black outfit with hat, coat, short pants, white leggings, and black buckle shoes. The Indian wears buckskin, holds a long rifle, and leans on a tall pile of furs. Between them is a chest from which the Dutch trader is taking cloth.

Sometimes trade and war go hand in hand. The "War Party" (702) has warriors all in buckskin. There is a painted post with hatchets buried in it, but the time period or event is not clear. "The Captive with his First Pair of Seneca Moccasins" (1022) confirms the fact that the Haudenosaunee captured and adopted other Indians during war to replace lost relatives.

Sullivan's campaign, a series of four paintings, presents the visual image of that disastrous time during the American Revolution. "Soldiers Shooting" (698) shows the soldiers shooting at the Seneca as they run away, and the bodies of several Seneca men lie on the ground. "Scout Murphy" (699) shows frontiersmen dressed in buckskin scalping a Seneca man as a soldier watches. "Houses and Cornfields Burning" (700) shows two Seneca log cabins on fire and a dead Seneca man and a wounded Seneca man trying to get up from the ground, as soldiers continue burning the village and cornfields. "Senecas Fleeing Burning Village" (701) shows the villages and cornfields burning as soldiers approach from the right side of the painting. At the left side, the Seneca men, women, and children are running for their lives. Older people and children are running into the woods as one warrior stands guard over them. They are dressed in both buckskin and cloth. A small child pulls on the hand of the warrior and points back at the burning village.

In "Progress," Ernest Smith presents a Haudenosaunee perspective on the invasion of Indian lands. Smith shows two groups of Haudenosaunee people dressed in leather, with their horses, high on a cliff overlooking a valley. In the valley is a white farm, with fenced fields; a steamboat chugs up the stream and a train passes over the stream on railroad trestles. In the background appears a large building with smoke pouring out the two tall smokestacks. There are very few trees left in the valley. The Haudenosaunee seem to be standing there contemplating these signs of "civilization."

Cultural Stories

It is impossible to talk about the Haudenosaunee way of life without telling stories. Vast volumes of Haudenosaunee stories have been collected.[25] Of the 240 Ernest Smith paintings at least half are illustrations of Haudenosaunee stories. These oral traditions explain how the world came to be, why nature is the way it is, how human beings should behave, and some are entertainment. The paintings of stories relating to the Thanksgiving Address, the three epic narratives, and corn have been included in those sections of this chapter. It would take a volume many times this size to discuss all the paintings and their stories.[26]

Ernest Smith illustrated the essential elements of Haudenosaunee world view, the three epic narratives, the importance of corn in the Haudenosaunee way of life, stories, ceremonies, and historic interaction. The values of cherishing children, seeing each part of nature as alive and having a spirit, of conservation, of extended family, and communal work are presented by Ernest Smith.

The combinations of buckskin/pottery/bark longhouse are significant in the pre-contact paintings of the Creation and the Great Law of Peace. The cloth clothing and plank board longhouse in "Handsome Lake Preaching" indicate post-contact time period.

Ernest Smith's paintings of ceremonial events include a mixture of buckskin and cloth clothing, corn pounders, splint baskets and iron kettles. This combination of pre and post-contact elements within the ceremonies was his way of showing that the ancient ceremonies continued but selective adoption meant that cloth and iron had been incorporated into the lifestyle. He painted the ceremonial events as they actually were in 1935. Today, these ceremonies with people wearing the same cloth ceremonial dress and following the same customs are continued in Haudenosaunee longhouses. Thus, we find selective adoption and continuity within this dynamic culture.

The thematic focus on corn was found throughout the paintings. Corn pounders appear in the background of many paintings, the origin of corn, the concept of the Three Sisters, ceremonies, planting, harvesting, and preparing of corn are all illustrated.

Continuity of Haudenosaunee World View in Contemporary Artistic Expression

The Ernest Smith paintings do not suggest that in 1935 Haudenosaunee people were dressing in blue jeans or dresses, as most people were in this country. One has to wonder if this omission was influenced by the still prevailing attitude of that era that Indians were a vanishing race, and research efforts were intended to capture the past. Perhaps Arthur Parker was trying to capture the past and had Ernest Smith illustrate only the ceremonial dress for the contemporary. In any event, students must have access to actual photographs from the time period so that they don't think the Haudenosaunee are still wearing buckskin and ceremonial dress on a daily basis. Haudenosaunee people do wear either buckskin clothing or the cloth clothing style illustrated by Smith, but only for ceremonial events.

Showing actual photographs of Jesse Cornplanter in traditional clothes, and Arthur and Ely Parker, who wore very dignified modern suits of the time period, would complement the paintings. Jesse Cornplanter was very proud of his traditional outfit and posed for photographers.[27]

Fred Wolcott's *Onondaga, Portrait of a Native People* photographed the Onondaga people from 1905–17. He captured the complexity of the time period, when dramatic changes were taking place. These photographs show that some Onondagas were financially well off and dressed quite well and others were very poor. The Wolcott photographs show Onondaga boys with bicycles, the baseball team, and a church.

Wolcott documented the continuity of the traditional culture in

> a display of wampum, the work of a resident lacrosse stick maker, the scene at a condolence ceremony, and a centuries-old farming format.[28]

William Fenton's "Tonawanda Longhouse Ceremonies: Ninety Years After Lewis Henry Morgan" documents the continuity of the longhouse ceremonies and provides photographs of Tonawanda in 1941. The Seneca people are dressed in the clothing of the time period, and they are pounding corn with a mortar and pestle for the Planting ceremony.[29]

It is absolutely necessary to use contemporary art and photographs in the classroom so that students will know that the Haudenosaunee are contemporary people who live in the modern world and continue their ancient world view. *Wisdom Keepers* provides interviews and photographs of a contemporary Onondaga Chief, Faithkeeper, and Clan Mother.[30] John Fadden's (Mohawk) drawings are well known and widely used in current publications such as *Keepers of the Earth*.[31] The Seneca Nation Museum has displays of Stan Hill carvings of the Three Sisters, and Carson Waterman's paintings, which combine the old and new. One of the Ernest Smith themes, that all of nature is alive and has a spirit which is illustrated with human faces in trees, with the corn, etc., can be found in paintings by contemporary Haudenosaunee artists.

Unbroken Circles, Traditional Arts of Contemporary Woodland Peoples (Dixon) pulls together ancient and the modern art forms illustrating how the ancient stories and values continue to be expressed by contemporary artists. From the moose antler carvings of the Three Sisters by Stan Hill to intricate corn husk dolls by Pam Brown and Gail General, the world view remains intact.

> Pam and Gail believe that the art of cornhusk doll-making links us to our past, through the corn, as part of the Three Sisters, our life staples. The cornhusks are fashioned into dolls that depict Iroquoian music, dance, traditional crafts, storytelling, and the lives of our ancestors. All aspects of Iroquoian life with the exception of ceremonial rites are depicted in the dolls.[32]

Cornhusk dolls do not have facial features. This allows children to "project his or her own emotions into the doll and imagine its expression and mood changes." Cole explains. As in almost every aspect of Haudenosaunee life there is a story that explains why the corn husk dolls have no face.

> At one time cornhusk people had very beautiful faces. The Creator placed them on earth to be companions to children. Their task was to play with and entertain little children. Pretty soon though, the cornhusk people became obsessed with their own beauty. They soon forgot their appointed task and spent most of their time staring at their reflections in calm pools of water. The children began to complain that the cornhusk people would not play with them. The Creator, disturbed at this, took away the faces of cornhusk people and made them into dolls.[33]

Unbroken Circles provides photographs of contemporary people in modern and traditional dress and many excellent works of art.

A special issue of *Faces, The Magazine About People,* September 1990, presents a mixed medium of art and time periods. A story by Jesse Cornplanter follows an interview with the contemporary Tadodaho, Leon Shenandoah. Several Ernest Smith paintings and a photograph of Julia Scrogg (1910) from Tonawanda with a corn mortar and pestle illustrate an article. She wears a long calico dress and a shawl over her head and shoulders. Included are photographs of sculptured carvings of Corn Spirit, Bean Spirit, and Squash Spirit by Stan Hill. A photograph of Tadodaho, and a faithkeeper in traditional clothes, and the Iroquois Nationals lacrosse team shows the continuity of the old ways within the context of a modern day sporting event. The only criticism of the issue is that they included the False Face medicine society.

Contemporary photographs and history appear in "The Fire That Never Dies," an article that deals with the continuity of the old and the selective adoption from the modern world.[34] This article contains the only contemporary published photograph from inside the longhouse. It took the the author two years to get permission to take photographs inside the longhouse.

Indian Corn of the America, Gift to the World (Barreiro 1989) provides information on the thematic focus on corn throughout the Americas. Photographs of contemporary Haudenosaunee and Mayan people in the cornfields shows the contemporary connection to corn. One photograph shows a grandmother and a young man, possibly a grandson, pounding corn with a mortar and pestle at Tonawanda in 1986. They wear blue jeans and sweat shirt or tee shirt and a little girl plays in the background.

The Ernest Smith paintings continue to be important within the Haudenosaunee community and can be found not only at Rochester Museum, but at the Tonawanda Community House, the Seneca-Iroquois National Museum on the Allegany Indian Reservation in Salamanca, New York, and in private homes. Some families have copies of prints published by the Canadian Indian Marketing Services, and some families have original paintings or sketches by Ernest Smith. Note cards of four of his sketches were available until quite recently. There are many publications where his works have been printed (see Appendix). A wall mural of a western scene with covered wagons and western Indians was painted by him on a bar wall during his later years when alcohol became a problem in his life.

This chapter has examined elements of Haudenosaunee culture as illustrated by Ernest Smith. The dynamic nature of the culture was examined by analyzing how Ernest Smith's paintings provide evidence of Haudenosaunee culture during the pre-contact time period, the post-contact period as Europeans arrived, and continuity of the culture up to the 1940s. The Haudenosaunee world view of being connected and interdependent with nature pervades the paintings. The thematic focus on corn can be followed from the "Sky Woman" painting of Ernest Smith, which illustrates the Creation story, to ceremonies, planting, harvesting, preservation, and preparation of corn for eating. The paintings show the healthy diet of the Haudenosaunee, which was a mixture of foods from fishing, hunting, gathering, and agriculture. The photograph of the Grandmother and young man pounding corn in 1986 provides a contemporary visual image of the continuity of corn in Haudenosaunee way of life.

The Ernest Smith paintings provide a visual image of Haudenosaunee culture that could enhance any unit of study on the Haudenosaunee. Throughout the paintings and contemporary photographs we find evidence of selective adoption and continuity within this dynamic culture. There are many resources for contemporary photographs and art that should become a standard part of every curriculum which seeks to provide another perspective on diverse cultures.

Chapter 9

The Contemporary Role of Corn in Haudenosaunee Culture

World view, that feeling of rootedness, of belonging to an ancient tradition, that "culturally-specific paradigm for viewing the universe," exists within the Elders' knowledge and the community's enactment of those beliefs.[1] In looking at the continuity of world view within the contemporary Haudenosaunee world, it is important to focus on these issues: Does corn continue to be culturally significant? Does corn continue to be a meaningful symbol and method of enacting world view in everyday living? Did the interviews in the Haudenosaunee communities confirm, modify, or negate the world view identified in the literature documentation and the Ernest Smith paintings? What evidence is there of selective adoption, adaptation, and cultural continuity? How has the role of corn changed due to historical interaction with other cultures?

The first section of this chapter analyzes the contemporary role of corn from the interviews in Haudenosaunee communities to find out if corn continues to be culturally significant by examining how things have changed yet stayed the same, i.e., selective adoption and continuity as discussed in chapter 8. The second section examines a corn project implemented by a family, with a committee of Cornell and Haudenosaunee people, and Cornell scientific researchers to examine this corn project's role in connecting with Haudenosaunee communities.

Contemporary Role of Corn

The living history of a culture includes world view, way of life, and historical events, which provide an understanding of why their world is the way it is. Enormous changes have taken place in Haudenosaunee agriculture due to U.S. government policies, such as the loss of lands after the American Revolution, and Quaker attempts to make the Seneca into white farmers. Millions of acres of land were taken by the United States government during the treaty-making period. In these treaties the Haudenosaunee leaders reserved portions of lands, territories, for their people, which today are called Indian reservations.

In the early 1900s, Haudenosaunee agriculturalists continued to operate self-sufficient homesteads throughout their territory. They raised white corn, beans, squash, other vegetables, fruit orchards, and domesticated animals. Their cellars were lined with home grown and home canned foods, and their corn cribs were full of white corn. They sold food they had produced and their crafts for a living, although some had already joined the wage labor market.

After the depression and World War II, the introduction of cars made access to cities possible and people left the reservation to find work in the cities, partly because employment was scarce or nonexistent on the reservations.[2] Today, approximately one-half of all American Indians live in cities as a result of government policies such as boarding schools and relocation programs. The change from agriculture to urban living has affected all of rural America during the same time period. The United States Department of Agriculture reports that thirty-nine percent of the U.S. population lived on farms in 1900, but by 1990 only two percent lived on farms.[3]

The pattern of Haudenosaunee agriculture changed from a communal village base to: 1) small individual family gardens, and 2) just a few families raising large quantities of corn to sell, a pattern that persists to the present day. Generally these gardens are small and each family raises enough corn to meet the family requirements. A survey of Haudenosaunee territory, conducted in 1990, found the following percentages of families planting small family gardens:[4]

Akwesasne	30–35%	Onondaga	20%
Oneida (N.Y.)	10%	Allegany	50%
Cattaraugus	60%	Tuscarora	15%
Tonawanda	30 families (%not provided)		

The Quintana survey found that white corn was generally planted in "small plots of four to sixteen rows," which is just enough to meet the family's needs. The traditional practices identified in this study included soaking the corn seed in "medicine," planting in hills or beds, planting when the oak leaves are "big as a red squirrel's foot," planting three days before the full moon, not planting when the east wind is blowing, cultivating after the first sprouting and again when the corn is two feet high, and intercropping corn, beans, and squash. Two cautions about women and planting state that a women should not plant during menustration, and a pregnant women should not walk through the corn fields.

Evidence of continued hill planting in the small gardens was found at Akwesasne and among the Oneida of Wisconsin and New York. The change to planting in rows seems to have taken place around the turn of the century but row planting was more evident in large fields, and in small family gardens hill planting and inter-cropping are preferred.

In each community, a few families have always raised large quantities of white corn. One of the themes found throughout the interviews was the awareness, and concern, that today *only* one or two families raise large quantities of corn on each Nation's lands. In some cases raising large fields of corn has diminished as the Elders got too old or journeyed to the Creator's land. Where this has happened, the people have to travel to other reservations to buy white corn. The older people can name five or six families that raised large quantities of corn when they were children, but today only one family, sometimes two, continue large-scale agriculture. Within the families who continue to raise corn there is a consistent pattern of agricultural knowledge being passed down from genera-tion to generation.[5] In fact, the community depends on specific families to raise surplus white corn for ceremonial use and to sell. In the traditional way the corn was not sold, but rather an ex-change was made, with the purchaser offering a gift to the grower.

Corn continues to be an essential part of Haudenosaunee ceremonies. The Three Sisters maintain their honored place as the Thanksgiving Address is spoken to open and close ceremonies and meetings. In ceremonies, corn appears in a variety of ways includ-ing corn soup, corn bread with beans, parched mush, and cracked corn. In a traditional Haudenosaunee wedding, the man's family must provide corn bread made with strawberries. The ceremonies are conducted in the Native language of the community.

In the ceremonies the people express their relationship with each other and Mother Earth. Several Elders strongly expressed their belief that raising a garden, planting the Three Sisters, was the way to understand and enact this connectedness with Mother Earth. These Elders said that as children they hated to garden, but their parents made them hoe and work in the garden. As adults, they have planted gardens, realizing the value of the connection between the earth, food, and their Elders. One Elder described this relationship as

> the joy of being able to walk over to the garden and pick peas, or carrots. . . . But you see, if you don't plant you miss that. If you're not into planting, you buy can goods, you don't buy the fresh vegetables. . . . If you're raised on can goods that's what you eat, and if you're raised that way you don't plant. If you eat canned goods you get all that salt into your system. Canned goods are not healthy. With the harvesting of Mother Earth's gifts, you are right out there. The very first are the leeks, then wild onions, cowslips, milkweed and then berries. You know you watch as these gifts appear.[6]

Comparisons have been made about the similarities between the cycle of corn and the pregnancy cycle. Each requires ceremonies and nurturing by human beings. A Maya spiritual leader shared this belief with a Mohawk woman:

> When a new baby is born the people in the home are supposed to be very happy and talk to each other like you want the baby to talk. Teach the new baby by your actions. In a home where there's a new baby you don't want any loud noises or angry people, or bad vibes, the same, he said, is with the corn because people used to hang it in their homes. . . . They said when corn is hanging in your kitchen it's the same as when there's a new baby in the house. You have to talk nice to each other. You have to be respectful of one another. You can't use angry language or bad words because the corn can hear you. If it hears all this negativity it won't grow for you the next year.[7]

Throughout the interviews, corn emerged as an element of the spiritual, physical, and emotional health of human beings.

Several people who plant small family gardens with five or six rows of corn mentioned problems they had with raccoons, groundhogs, and deer. They were upset because they lost most of their corn crop to these animals. The Elders said that long ago there were two views of this problem: first, that such huge fields of corn were planted that losing the first five or six rows to animals was not a hardship, and, secondly, planting a larger quantity of corn takes care of the problem when you understand that humans must be willing to share part of the crop with the animals. Sharing with the animals is part of nature's cycle, as expressed by Ernest Smith in "The Chipmunk," (737) which shows a chipmunk eating from a wooden bowl.

Some families plant white corn for family use and donate corn to the longhouse for ceremonies. Other families raise large quantities of corn to sell to others who prepare and sell their favorite corn dishes at community events. The people who prepare the corn to sell are not necessarily the same people who grow corn.

In some communities, on any given day, a good corn soup maker can hang out her/his sign, "Corn Soup Today," and people will stop by to buy corn soup. One early spring day at Tonawanda I saw this sign, "Wild Onion Soup at the Longhouse Today," and there were cars parked outside the longhouse. The community depends on certain people with the skills to make good corn soup or wild onion soup to prepare these foods and put out the sign. Some of the Elders interviewed mentioned that they grew up eating corn soup regularly at ceremonies, and in between ceremonies there was always a "Corn Soup Today" sign up; thus they ate corn soup on a regular basis. Almost everyone who has grown up in a Haudenosaunee community has learned to love corn soup either at home or at special community events such as social dances, lacrosse games, birthday parties,weddings, family events, and ball games. In fact, one such community event is the annual Fall Festival at the Cattaraugus Seneca Reservation in western New York state. It developed as a fair to display the products of Haudenosaunee farmers in the Iroquois Agricultural Society during the early 1900s. People come to this fair from long distances to enjoy corn syrup and renew family relationships. Today it has more of a carnival theme, but there continues to be a display of corn, beans, squash and other vegetables as part of the Fall Festival.

Within those families who raise corn and have the knowledge of preparation, family life centers around corn. These families enjoy corn soup, corn bread, parched mush, roast corn, and ogo'sa which is dried green corn. Preparation of the corn by boiling it with wood ashes is called washing the corn. The corn basket described by Parker continues to be the best equipment for washing the corn. Preparing corn requires wood stove ashes, a corn basket, some sort of corn grinder, a flat wide wooden paddle, and wooden spoons. Corn pounders are generally found at the "mudhouse" or cook house, which is a separate building next to the longhouse where corn is prepared for ceremonies. Today, for home use, a metal grinder that is turned by hand or an electric blender can be used. Making corn soup is a family event as children sort the corn, the mother or father takes care of the boiling water and ashes, and children use the corn basket to rinse the corn. If corn bread or parched mush is made the children take turns with the hand grinder. Maple syrup is used to sweeten the parched mush, which can be eaten with salt pork, or as a cereal with milk and sweetener.

Katsi Cook shared a family event surrounding corn and her family. Making corn bread is a two-day family event because the corn must be washed, dried, and then ground into flour. The grinding is generally done with a metal hand grinder attached to the kitchen table and it takes hours to accomplish the grinding task as family members take turns cranking the grinder. One day, after many complaints, she went to town and bought a fancy grinder and could process the dried corn into flour within minutes. After several months, she complained that her family wasn't in the kitchen as often and realized how much talking she did with her children while they ground the corn. " Well," she said, "I just put that fancy new machine in the bottom of the cupboard and attached the old metal grinder to the kitchen table and spent time talking with my children as we ground corn."

Utilization and appreciation of corn continues to be an important aspect of Haudenosaunee culture. A pamphlet, *Iroquois Recipes,* was published by the Seneca Bilingual Education Program at Allegany during the 1970s. This pamphlet provides recipes for parched mush (o'showe:h), corn bread (gagaehde'hdo'), hominy (cracked corn—o'nohda:h), dried corn (o'gosa'h), roast corn soup (o'nehda'), hull corn soup (ono':hgwa'), and a modern adoption, fried bread made with flour, also called ghost bread (ga:hgwagi:da:h).

Another booklet, *Iroquois Cookbook,* written and produced by Seneca women, was published in 1989 to raise funds for the Peter Doctor Memorial Indian Scholarship Foundation, which supports young Haudenosaunee people attending college. People donated recipes from Allegany, Onondaga, Tonawanda, and Tuscarora. The cover of the booklet, an illustration of the Three Sisters, was selected because it

> represents the "three sisters"—corn, beans, and squash, depended on by earliest Iroquois for sustenance. . . . We have

selected it for the cover as appropriate since so many of the recipes still include the corn, beans, and squash.[8]

The traditional Haudenosaunee corn foods listed above are included in this cookbook, plus ways to cook: pumpkins, wild onions, leeks, puff balls, squirrel, milkweed greens, wild mushrooms, rabbit, raccoon, muskrat, venison, wild grapes, sassafras tea, wild elderberries, and wild huckleberries. This little cookbook provides evidence that in 1989 the Haudenosaunee continued to have the cultural knowledge and skill to utilize nature's gifts from fishing, hunting, gathering, and agriculture.

Corn Project With Cornell

The corn project is situated at Crows Hill Farm, about fifteen miles southeast of Ithaca, New York. It is an independent, non-profit project of the Indigenous Preservation Networking Center (IPNC), which has a Native American Board of Directors. The project began as a way for a family to sustain the Haudenosaunee agricultural cycle and indigenous values while living away from the reservation. As people from different Haudenosaunee communities and Ithaca visited the farm to ask about the corn, the family garden developed into a project with the purpose of strengthening the white corn seed. Officially organized in 1985, the project has a committee composed of Haudenosaunee agriculturalists and scholars, and Cornell scientists, including Jane Mt. Pleasant, Ph.D. from Tuscarora, and Jorge Quintana, Ph.D. from Nicaragua.

Barreiro describes the value of raising corn, which he says emerged from listening to many people talk about corn in their own lives:

> There are concentric circles around it (planting corn) in terms of whether you're looking at the language of it, or individual stories people have about it, in terms of their own family, what they know about it, what they've done, what they remember of the corn, and what it means to them. Almost everybody who's done these activities as a child never forgets it. It always stays a part of their identity.[9]

The project came about at a time when people were expressing concern about the white corn seed, which was producing shorter, less abundant ears and developing a fungus type of smut. An elder told Barreiro, the biggest problem is that people don't share the seed anymore, don't exchange seed, so the seed starts to go down, to lose vitality, (it) kind of inbreeds.

The people participating in the corn project were able to gather seed from four of the Haudenosaunee reservations with a total of five seed sources. Barreiro described the corn project participants' attitude toward the corn, saying they felt corn was to be

treated as a gift to us, not something we own. We didn't want to do something with it that would be far afield from how people generally feel about it in the community culture.

The project seeks to integrate traditional knowledge on planting and preparation of corn foods, as well as to explore possible marketing strategies, within environmentally sound agricultural practices. The researchers used a process called natural selection, the same process used by American Indian corn growers in antiquity to select ears for a strong stalk, a good sized ear, resistance to disease, and overall strength of the corn. Each year, using these ancient methods, the corn seed has improved until now it is almost free of smut, stands up well to the weather, has large ears, and its productivity has increased. The experiments continue to strengthen the seed. The goals of the project are:

1. to promote the preservation of indigenous scientific and cultural values and knowledge;

2. to incorporate this knowledge into modern systems of ecologically sound sustainable production;

3. to encourage the socioeconomic development of indigenous communities and nations;

4. to enable cultural and scientific interchange and understanding among people with different ethnic backgrounds;

5. to transmit information to individual members of the communities affected and to assist as requested in the application of the information.[10]

The corn project survey of Haudenosaunee reservations found that vast amounts of Haudenosaunee land has been leased to non-Indian farmers for a number of years. The survey confirmed what interviews with the Elders had revealed, that there are many small family gardeners among the Haudenosaunee who maintain the strong agricultural traditions that have been handed down from generation to generation within specific families.[11]

Mt. Pleasant has pointed out that no-till planting, ridge tillage, and intercropping are ancient Haudenosaunee agricultural techniques, which researchers at Cornell are now discovering to be

> sound agronomic principles, and are the same principles that we still use in 1988 for growing good corn.[12]

Responding to requests from American Indian communities in Central America and the Caribbean has also become a part of the work. Researchers from this project have visited projects among the Maya in Guatemala, the Caribs of Dominica, and the Miskito and Nahua Indians of Nicaragua. This work with other indigenous peoples has provided a way to share and exchange agricultural knowledge.

The project has encouraged a two-way communication with Haudenosaunee communities. The people from the project go out to the communities and people from the communities travel to the project. White corn seed from this project is now planted in several Haudenosaunee communities, and the corn project is well known throughout Haudenosaunee lands. At harvest time, over a period of two weeks, people young and old arrive from various Haudenosaunee communities and the Ithaca area to help with the harvest. An Elder sings a song before the harvest begins, corn stories are told during the picking, the young people make corn husk dolls, and discussions are held about the significance of finding a "red ear of corn," and as each person

> pulls back the husk, it's like seeing a brand new baby the way everybody reacts. "Look at this one, here's one red kernel among all the white."[13]

At the end of the day, the exhausted but exhilarated workers devour parched corn mush with maple syrup and hot coffee, and tell jokes causing much laughter and joy. As the workers huddle around the woodstove to get warm the discussion turns to the Grandfather corn hanging from the rafters in the shed.

This project has received gifts of various seeds from a variety of people. One of those gifts was a partial ear of Delaware Grandfather Corn, also known as pod, husk, or original corn because each kernel has its own husk or glume in addition to the regular husk.

This corn grows much taller than regular varieties and has a sturdy purple-streaked stalk. This unique corn has provided ears of corn with kernels of red, white, yellow, deep navy, blue, purple, and calico. It also has reproduced the original deep purple ear of corn, as well as beige ears with individual husks surrounding each kernel. Some ears of this corn have the individual husks and some do not. The Grandfather Corn appears to be a genetic seed bank. The Grandfather Corn has strength.

The original donor of the Grandfather Corn was a man passing through the area who asked that the seed be planted to preserve it. The impression was that he obtained the corn from a Seneca farmer, but that information didn't turn out to be correct, so the origin of this particular little ear of Grandfather Corn is unknown at this time. The only other source of this corn is in Wisconsin where an Oneida construction worker in Ohio found a sealed clay pot containing an ear of the Grandfather Corn which was then planted on the Oneida reservation in Wisconsin. In asking about the pod corn around Haudenosaunee Territory the following information surfaced:

> There are a couple of old people who have mentioned knowing about it in the early 1920s and 1930s. We've gotten one reference to it at Grand River (Six Nations Reserve in Canada) during the Moon Dance. It is used to dip in water and anoint the young women who were dancing for the first time. . . . A Mohawk basketmaker from Akwesasne remembered having seen it as a young girl and mentioned there was a Mohawk word for it, but couldn't remember the word, but she did remember seeing it in people's gardens. An Onondaga faithkeeper said "This is really significant. This is like sitting here with an old, old timer." And you could tell he really felt that way, and other people have too. Even if they don't know anything about it, they know there's something to it because there's so much reverence for the regular corn in the culture and what it represents.[14]

The only reference located for the Mohawk word which the Elder couldn't remember was oo'na' (pod corn) and ona'o we (sacred corn or original corn).[15]

In other American Indian cultures knowledge of the pod corn was found among the Guaicuru Indians of Brazil, the Guarnay of Paraguay, and the Quichua women of Bolivia. The Humpbacked Flute Player of the Andean highlands travelled throughout South and Central America with a bag of medicines and his flute. One of his medicines was pod corn. The husks from the individual kernels were used to cure problems related to the chest and breathing such as tuberculosis, asthma, and bronchitis.[16]

Archeologists and researchers have conducted studies for years in the attempt to find the origins and antiquity of corn. Archeologists found pod corn in the Bat Cave of New Mexico that was radiocarbon dated as 5,600 years old. Fossil pollen of pod corn was found beneath Mexico City, which confirmed that in the Valley of Mexico corn was grown at least eighty thousand years ago.[17] Pod corn was found in the Tehuacan valley of Mexico that was radiocarbon dated at 5,000 B.C.[18]

Mangelsdorf developed the Tripartite theory which says the "ancestor of cultivated corn was a form of pod corn."[19] The second and third sections of the theory discuss unknown wild grasses that may be the ancestor of pod corn, but to date the actual origins of corn and the exact geographical location remain a mystery.

Evidence of pod corn was found in the ancient pottery from the Peruvian highlands. But Mangelsdorf maintains the indigenous stories of the origins of corn "are fantastic in the extreme and merit little credence," and he says the Peruvians had no legends about the origins of corn.[20] On the other hand, they probably wouldn't have shared their knowledge with this man who had little respect for their beliefs. I recently asked a Quechua family from the highlands of Peru about Mangelsdorf's assertion that they did not have stories about the origins of corn, and they laughed, saying of course they had stories. Having planted some of the pod corn from the Cornell project here in Oneida territory in Wisconsin, I recently had an opportunity to show my ear of pod corn to indigenous people from Ecuador and their response was "mucho mama."

The origins of the gift of the Grandfather Corn remains a mystery, but its strength and genetic diversity are appreciated by those who have walked among the tall stalks and have witnessed the genetic diversity the Grandfather corn returns to the people who care for it.

Thus, we find the corn project, which began on a small family-based scale, has grown into an extended family/community-based project that interacts with the Haudenosaunee communities and Cornell researchers to share agricultural knowledge from both worlds. It is an interaction that utilizes selective adoption (in the fall of 1991 they tried out a one-row corn picker) and continuity in maintaining the role of corn and community by having people from the Haudenosaunee communities and non-Indians from many walks of life participate in the harvest.

Summary

The significant themes that emerged from the interviews, literature documentation, and Ernest Smith paintings provide a coherent, continuous pattern of the importance of corn in Haudenosaunee
world view and way of life, past and present. The Haudenosaunee
world view continues to be expressed by the Haudenosaunee communities. The Thanksgiving Address and the ceremonial cycle continue to express the sense of connection, the relationship with earth,
with nature, which is maintained by raising corn on a small family
garden and sharing corn as a community. A consistent theme is the
appreciation of corn as gift from the Creator. The "culturally-specific
paradigm" interweaves world view, the value of raising corn within
an extended family network, and the enjoyment of eating corn
foods. Thus, we find spiritual, emotional, and physical expression
of the relationship with corn.

The people utilize corn in their way of life by raising corn,
participating in the ceremonial cycle, and eating corn foods at ceremonies and special community events. Families pass their knowledge of raising corn from generation to generation. People expressed
an intrinsic value of raising corn for the family to provide the
closeness that comes from working together to plant, take care of,
harvest, prepare, and eat the corn. Cultural beliefs about when to
plant corn, planting in hills, when to cultivate, and intercropping
are evident in the interviews and the survey conducted as part of
the Corn Project.

The interaction of cultures has had a tremendous impact on
Haudenosaunee agriculture. It changed from large-scale communal
agriculture to small individual family gardens. The impact of cars,
machinery, and electricity changed the face of agriculture across
this country. Within the Haudenosaunee communities the change
resulted in only five or six families continuing large-scale production of corn. In some of the small-scale family gardens corn, beans,
and squash continued to be planted in hills. However, at the present
time it is rare to find evidence of hill planting except occasionally
in a family garden. The number of Haudenosaunee farmers producing corn on a large scale has diminished to one or perhaps two
farmers in a community.

The dynamic nature of the role of corn can be found in the
selective adoption of machinery from the modern world. Corn pounders continue to be used at the longhouse for ceremonial preparation
of corn. However, at the family level a hand cranked metal grinder
or electric blender is used for this task. This can be compared to

Ernest Smith's time, the 1930s, when he illustrated the Woman's Dance for the corn during the Seed Ceremony, and the women wore cloth dresses and a cast iron kettle was used for the corn soup. In both instances, the traditional ceremony and grinding corn, selective adoption has taken place within the context of continuing the traditions involving corn.

The continuity of the role of corn in the Haudenosaunee way of life was found within the ceremonial cycle, planting white corn, people's appreciation of the gift of corn, and the many events at which people continue to eat corn foods. The Corn Project survey found that on the average, one-third of the families throughout Haudenosaunee territory raise corn in family gardens. The knowledge surrounding corn continues to be culturally significant and passed down from generation to generation. Elders express a concern that children who are raised on canned foods from the grocery store don't have that same reverence, respect, and connection to the earth as the children who take part in raising the family garden. Two recent Iroquois cookbooks show that the Haudenosaunee continue to prepare foods obtained from hunting, fishing, gathering, and agriculture.

The Corn Project of the IPNC began as an individual family effort and grew to include the extended family of Haudenosaunee communities, the Ithaca community, and Cornell scientists. It continues to be a small-scale project dedicated to preserving traditional agricultural knowledge. The gift of the Grandfather Corn, or pod corn, continues to be a blessing and a mystery.

The varieties of Iroquois corn seed and bean seed have been slowly collecting at the Cornell project. As travellers pass through they share their seeds, thereby dispersing varieties throughout Haudenosaunee country. And when the people involved in this project travel to Haudenosaunee communities, there are people willing and eager to share their seed varieties. We have seen the seeds grown once again that we thought were lost. Recently, I obtained from a friend some Cornplanter beans, which are said to have come from Cornplanter's reservation. That reservation no longer exists because it was lost to the building of Kinzua dam in the 1960s. As I finish writing this book, the cornplanter beans are flourishing in my garden along with three other varieties of beans, four experiments with the pod corn, twenty hills of the Three Sisters, and rows of white corn. In my own experiments this year, I found that the twenty hills of Three Sisters grew taller, are greener, and have larger ears of corn than the corn planted in rows, which are shorter, have a yellowish-green color, and have not only fewer but shorter ears of corn! The scientific experiments continue as Haudenosaunee people continue to exchange seeds.

The literature documentation on the Haudenosaunee, the Ernest Smith paintings, and the interviews with Elders combine to provide an understanding of the complex, interconnected elements of the Haudenosaunee culturally specific paradigm.

In a recent interview with Katsi Cook (Mohawk) we compared the similarity between raising corn and making a corn basket. Each splint of the basket must be woven in correctly to produce the whole basket, which has a specific purpose. Modern technology has not produced a utensil that can replace the corn basket. So it is with corn, which has spiritual, emotional, and physical aspects woven together in a culturally specific paradigm, a way of life.

Appendix 1

Seneca Thanksgiving Address by Chief Corbett Sundown, Tonawanda Seneca, 1959

The People

And now we are gathered in a group. And this is what the Sky Dwellers (The Four Beings) did: they told us that we should always have love, we who move about on the earth. And this will always be first when people come to gather, the people who move about on the earth. It is the way it begins when two people meet: they first have the obligation to be grateful that they are happy. (Lit. "that they are thinking well," with reference to both mental and physical health.) They greet (or "are thankful for") each other and after that they take up the matter with which just they two are concerned.

And this what Our Creator (lit. "he fashioned our lives") did: he decided, "The people moving about on the earth will simply (i.e., it's all that will be required of them) come to express their gratitude." And that is the obligation of those of us who are gathered: that we continue to be grateful.

Chafe, Wallace L. *Seneca Thanksgiving Rituals*. Smithsonian Institution, Bureau of American Ethnology, Bulletin 183: United States Government Printing Office, Washington, D.C. 1961: 17–45.

This, too, is the way things are: we have not heard of any unfortunate occurrence that there might be in the community. And the way things are, there are people lying here and there, held down by illness; and even that, certainly, is the responsibility of the Creator (i.e., it is for him to decided whether or not they will recover). And therefore let there be gratitude; we are always going to be grateful, we who remain, we who can claim to be happy. And give it your thought: the first thing for us to do is to be thankful for each other. And our minds will continue to be so.

The Earth

And now this is what Our Creator did: he decided, "I shall establish the earth, on which the people will move about. The new people, too, will be taking their places on the earth. And there will be a relationship when they want to refer to the earth: they will always say 'our mother, who supports our feet.'"

And it is true: we are using it every day and every night; we are moving about on the earth. And we are also obtaining from the earth the things that bring us happiness. And therefore let there be gratitude, for we believe that she has indeed done all that she was obligated to do, the responsibility that he assigned her, our mother, who supports our feet. And give it your thought, that we may do it properly: we now give thanks for that which supports our feet. And our minds will continue to be so.

The Plants

And now this what the Creator did. He decided, "There will be plants growing on the earth. Indeed, all of them will have names, as many plants as will be growing on the earth. At a certain time they will emerge from the earth and mature of their own accord. They will be available in abundance as medicines to the people moving about on the earth." That is what he intended. And it is true: we have been using them up the present time, the medicines which the Creator made. He decided that it would be thus: that people would be obtaining them from the earth, where the medicines would be distributed. And this what the Creator did: he decided, "Illness will overtake the people moving about on the earth, and these will always be there for their assistance." And he left on the earth all the different medicines to assist us in the future.

And this too, the Creator did. With regard to the plants growing on the earth he decided, "There will be a certain plant growing on the earth," he decided, "There will be a certain plant on which berries will always hang at a certain time. I shall then cause them to remember me, the people moving about on the earth. They will always express their gratitude when they see the berries hanging above the earth." And the Sky Dwellers called them shes?a:h (a term reserved for the ceremonially important wild strawberry). But we who move about on the earth shall always call them jistota?shae? (the generic word for strawberry, wild or cultivated, lit. "embers attached to it"). And it is true: we see them when the wind becomes warm again on the earth; the strawberries are indeed hanging there. And it is also true that we use them, that we drink the berry water (the ceremonial mixture of strawberries and water). For this is what he did: he decided, "They will always bring them to their meeting place and give thanks, all the people, as many as remain. They will be thankful when the see the berries hanging." That is what he did. And it is true: it comes to pass. When in the course of things it becomes warm again on the earth, we are thankful for everything. And give it your thought, that with one mind (lit. "we establish our minds as one") we may give thanks for all the plants, our medicines. And our minds will continue to be so.

The Water

And this is what the Creator did: he decided, "There will be springs on the earth. And there will be brooks (lit. "veins, arteries") on the earth as well; rivers will flow, and will pass by under the earth. And there will also be ponds and lakes. They will work hand in hand, the way I fashion them on the earth. And moisture will continue to fall." And it is true: fresh water is available in abundance to us who move about on the earth. And, in fact, to all those things which he provided for our contentment, fresh water is abundantly available too. And it is true: we have been using it up to the present time. It is the first thing we use when we arise each new time. When the new day dawns again, the first thing we use is water. And let there indeed be gratitude. It is coming to pass as Our Creator intended. And give it your thought, that we do it properly: we now give thanks for the springs, the brooks, the flowing rivers, and the ponds and lakes. And our minds will continue to be so.

The Trees

And now this is what the Creator did. He decided, "There will be forests growing on the earth. Indeed, the growing forests will be of assistance to the people moving about on the earth." He decided, "There will always be a certain period when the wind will become warm, and a certain length of time, also, when it will become cold. And the forests growing on the earth will provide heat for them." That is what the Creator intended. And it is true: it continues unchanged up to the present time. A few of us are using them for heat, the forests growing on the earth. And this also he did: he made them medicines as well, the trees growing on the earth. He decided, "They can also be available as medicines to the people moving about on the earth." And he even did this as well: he decided, "Again, there will be a certain tree which I shall cause to remind the people moving about to think of me. The maples will stand on the earth, and the sweet liquid will drip from them. Each time when the earth becomes warm, then the sap will flow and they will be grateful for their happiness. When the time arrives again, they will attend to the maples standing there." And for those people who take notice of it, it continues unchanged: they do indeed tap them and store the sugar. For he decided that it would be available in abundance to the people moving about on the earth. And it is true: it continues unchanged up to the present time; we are still using it. And therefore again let there be gratitude that it all still continues as the Creator planned it. And give it your thought, that we may do it properly: we now give thanks for the forests growing on the earth. And our minds will continue to be so.

The Animals

And now this is what Our Creator did: he decided, "I shall now establish various animals to run about on the earth. Indeed, they will always be a source of amusement for those who are called warriors, whose bodies are strong." He decided to provide the warriors, whose bodies are strong, with the animals running about, to be a source of amusement for them. "And they will be available as food to the people moving about on the earth." And up to the present time we have indeed seen the small animals running about along the edges of the forests, and within the forests as well. And at the present time we even catch glimpses of the large animals again. There were in fact a number of years during which we no longer saw the large animals. But now at the present time we again see the large animals running about, and at the present time they are actually available to us again as food. And we are using them as Our Creator intended. And therefore let there be gratitude that it all does still continue as he intended. And give it your thought, that we may do it properly: we now give thanks for the animals running about. And our minds will continue to be so.

The Birds

And this is what Our Creator did. He decided, "I shall establish various creatures that will spread their wings from just above the earth to as far upward as they can go. And they too will be called animals. They will begin just above the earth, and will go all the way into the clouds. And they too all have names, the birds with outspread wings." And with respect to the small birds he decided, "There will be a certain period when they will stir, and they will turn back, going back to where it is warm. And it will become warm again on the earth, and they will return. With all their voices they will sing once more their beautiful songs. And it will lift the minds of all who remain when the small birds return." And he arranged as well that they are available to us as food, the birds with outspread wings. It is true: we are using them too, the birds with outspread wings. They are available to us as food. And we believe that they too are all carrying out their responsibility. They all, as I said, have names, according to their types. And give it your thought, that we may do it properly: we now give thanks for the birds with outspread wings. And our minds will continue to be so.

The "Sisters"

And now this what Our Creator did. It was indeed at this time that he thought, "I shall leave them on the earth, and the people moving about will then take care of themselves. People will put them in the earth, they will mature of their own accord, people will harvest them and be happy." And up to the present time we have indeed seen them. When they emerge from the earth we see them. They bring us contentment. They come again with the change of the wind. And they strengthen our breath. And when the Good Message came we were advised that they too should always be included in the ceremonies, in the Four Rituals. Those who take care of them every day asked, too, that they be sisters. And at that time there arose a relationship between them: we shall say "the Sisters, our sustenance" when we want to refer to them. And it is true: we are content up to the present time, for we see them growing. And give it your thought, that we may do it properly: we now give thanks for the Sisters, our sustenance. And our minds will continue to be so.

The Wind

And now this is what Our Creator did: he decided, "Now it can't always be just this way," And this, in fact, is what he decided. "There must be wind, and it will strengthen (by providing them with air to breathe) the people moving about whom I left on the earth. And in the west he made the thing that is covered by a veil; slowly it moves and revolves. There the wind is formed, and we are happy. It indeed strengthens our breath, for us who move about on the earth. And the wind is just the strength for us to be content with it and be happy. But the Sky Dwellers told us: they said, "We believe that your kinsmen will see that in future days it may happen that it will be beyond our control. It is the most important thing for us to watch. It may become strong in its revolving, and we believe that it will scrape off everything on the earth. The wind may become strong, we believe, and bring harm to the people moving about." That is what they said. And indeed up to the present time we can attest to it: the way it occurs, it destroys their homes. From time to time it is destructive, for the wind can become strong. But as for us, we are content, for no matter how strong the wind has been we have been happy. And give it your thought, that we may do it properly: we give thanks for the thing that is covered by a veil, where the wind is formed. And our minds will continue to be so.

The Thunderers

And now this is what Our Creator did: he decided, "I shall have helpers who will live in the west. They will come from that direction and will move about among the clouds, carrying fresh water." They will sprinkle all the gardens which he provided, which grow of their own accord on the earth. And he decided, "There will be a relationship when people want to refer to them: they will say 'our grandparents, the Thunderers.' That is what they will do." And he left them in the west; they will always come from that direction. And truly they will always be of such a strength that the people, their grandchildren, who move about will be content with them. And they are performing their obligation, moving about all through the summer among the clouds, making fresh water, rivers, ponds, and lakes. And give it your thought, that we may do it properly: we now give thanks for them, our grandparents, the Thunderers. And our minds will continue to be so.

The Sun

And now this what Our Creator did: he decided, "There will be a sky above the heads of the people moving about. I must have a helper in the sky as well." And indeed he assigned him to be attached to the sky. There he will move about, and will cross the earth. He will always come from a certain direction, and will always go in a certain direction. And he also prescribed a relationship when we want to refer to it: we shall say "our elder brother, the sun." And it is true: he is carrying out his responsibility, attached there to the sky; there is beautiful daylight, and we are happy. And we believe that he too has done all that he was obligated to do; everything that he left to grow of its own accord is flourishing. He gave them the added responsibility of making it warm on the earth, so that everything he left to grow of its own accord would flourish. And we believe that he is performing his obligation up to the present time, the assignment he was given. And give it your thought, that we may do it properly: we give thanks for him, our elder brother, the sun. And our minds will continue to be so.

The Moon

And now this what Our Creator did: he decided, "There will be a certain period when the earth will be in shadow, as well as a certain period when it will be day." And indeed he saw well that the people moving about were taking care of themselves. And he decided, "They will rest. They will lay down their bodies and rest while it is in shadow." That is what he intended. "And perhaps it will happen that somewhere at a distance (from home) they will run into darkness. And I shall have another helper, another orb in the sky. People will say 'our grandmother, the moon.' That is how they will do it. It can be a sort of guide for their steps, providing them with light." And indeed it is a measure for us as we go along, we who move about on the earth. He decided, "The moon will change its form as it goes." They have called it "phases." And it is true: it is still a measure for us up to the present time, the way it is as we go along, we who move about on the earth. And we believe that they come from there too, that it continues unchanged: the little ones taking their places on the earth. (i.e., the cycle of reproduction is determined by the moon). They are here and they come from our mothers. And therefore we believe that she has done all that she was obligated to do, the assignment she was given. And now give it your thought, that we may do it properly: we now give thanks for her, our grandmother, the moon. And our minds will to be so.

The Stars

And now this what Our Creator did. He decided, "There will also be stars arrayed in the sky while it is dark." And he assigned to them certain things as well, the way it would continue to be. He decided, "They too will all have names, all the stars in the sky. And they too, in fact, will be indicators, to be used for measuring by the people moving about. If it happens that they run into darkness on their journey, they will use them, the people moving about. And indeed they will lift their faces to the stars and will be set straight. They will head back directly toward their homes." And up to the present time they have had an added responsibility. While it is dark they will cause moisture to fall on everything that he left to grow of its own accord on the earth. And truly they enjoy water throughout the night, everything that he left to grow of its own accord. It comes from the stars arrayed in the sky. And we believe that they are performing their obligations, the responsibility that they too have. And give it your thought, that we may do it properly: we now give thanks for them, the stars arrayed in the sky. And our minds will continue to be so.

The Four Beings

And now Our Creator decided, "I shall have the Four Beings as helpers to protect the people moving about on the earth." Indeed, he saw well that it was not possible for them alone, that they could not continue to move about alone. It was true: all sorts of things were going on on the earth where they would move about. It was inevitable that the people moving about on the earth would have accidents. The people moving about on the earth would have accidental things happen to them that would be beyond their control. And indeed we too can attest to it, we who move about on the earth: it will happen that people are involved in accidents that are beyond their control. It is the way with us who move about on the earth. And indeed they also have the added responsibility of keeping watch over those of his helpers called the Four Groups. (To keep in check the wind, the Thunderers, the sun, and the moon, who might otherwise bring destruction.) They will continue to look after us whom he left on the earth, and will bring us contentment. And we believe that they too are performing their obligation, the assignment they were given, those who are called the Four Beings, our protectors. And therefore let there be gratitude, for we believe that we are happy. Give it your thought, that with one mind we may now give thanks for his helpers, the Four Beings, our protectors. And our minds will continue to be so.

Handsome Lake

And now this is what Our Creator did. He did indeed decide it, and it must happen according to his will. Indeed he (Handsome Lake) was among us who moved about on the earth. Illness took hold of him, and he was confined to bed. For a number of years he lay helpless. And the way things were, he had to be thankful during the nights and the days, and he thought that there must be someone there who made all the things that he was seeing. And thereupon he repented everything, all the things he thought he had done wrong when he moved about on the earth. And indeed he was thankful each day for each new thing that he saw. And now it happened that the Creator saw well how the people on the earth were acting. It seemed that nowhere was there any longer any guidance for the minds of those who moved about. And now it happened that he sent his helpers to speak to our great one, whom we used to call Handsome Lake, when he moved about. They gave him the responsibility to tell us what we should do in the future. And for a number of years he told about the words of the Creator. And the way things went, he labored until he collapsed. And let there indeed be gratitude that from time to time now we again hear the words of the Creator. And therefore let there be gratitude that it is still continuing as he planned it. And give it your thought, that we may do it properly: we give thanks for him, whom we called Handsome Lake. And our minds will continue to be so.

The Creator

And now this is what Our Creator did. He decided, "I myself shall continue to dwell above the sky, and that is where those on the earth will end their thanksgiving. They will simply continue to have gratitude for everything they see that I created on the earth, and for everything they see that is growing." That is what he intended. "The people moving about on the earth will have love; they will simply be thankful. They will begin on the earth, giving thanks for all they see. They will carry it upward, ending where I dwell. I shall always be listening carefully to what they are saying, the people who move about. And indeed I shall always be watching carefully what they do, the people on the earth." And up to the present time, indeed, we people believe that we are happy. And therefore let there also be gratitude that we can claim to be happy. And give it your thought, that with one mind we may now give thanks for him, Our Creator. And our minds will continue to be so.

Epilogue

And that is all that I myself am able to do. What they (The Four Beings) did was to decide that a ritual of gratitude, as they called it, would always be observed in the future, when in the future people would gather. And that is all that I myself am able to do; that is all that I learned of the ritual which begins the ceremony. That is it.

Appendix 2

Long Opening Thanksgiving Address
by Enos Williams (1915–1983), Grand River, 1974

(This version was given in the Cayuga Language)

Prologue and Assembly

And now the time has come. The responsibility for this speech has been given to me. From me will issue all of our words, However far it may be possible. I ask that you please have patience in this matter.

And this is what he himself did, he who in the sky dwells, Our Creator. He decided, "I will create some people for myself that will move about upon the earth." And indeed we are moving about still, those of us who up to the present time still remain. And he gave us what we are to do before we take up anything else.

For he decided, "This is the way it will begin: They will thank me. And it need be only in their thoughts that they start in. And if they should meet one another as they move about, they will have their Thanksgiving Address. And also when people gather from time to time. They will have it before they put a ceremony through. They will be grateful that so many people are happy—Contented that their minds have been brought together."

From Foster, Michael, K. *From the Earth to Beyond the Sky: An Ethnographic Approach to Four Longhouse Iroquois Speech Events.* National Museums of Canada, Ottawa, 1974: 285–360.

213

And now today, this day, the time has come, for it is possible that our minds have come together. We are all happy. And also we are contented. So let there be gratitude. And it must be so, it must be apparent, that everyone in the families is well. For we have not heard of any family anywhere hit by unhappiness. It is true that some are lying in sickness, in distress. But there is nothing we can do about it. We ask only that it might be possible for everyone to be healthy, those of us moving about over the handiwork of our Creator. The responsibility lies with him alone. All that makes us happy was left here by him. And now this group of us is happy, for our minds have come together. So let it be our thought to begin by greeting one another, and so it will be in our minds.

The Earth

And now we will speak about what he has done. He decided, "I will create a world below the Sky World, And there they will move about, the people I create on the earth. And there is a way people will have to refer to it as related, This the earth: 'Our Mother, it is related to us, that which supports our feet.'" And it so still, It has come to pass that we are still moving about. That was his determination: he decided where we should be moving about. And also all that he left will be contributing to our happiness on the earth. And it is so still. It is possible that it comes from the earth, the happiness we are obtaining. For all this, therefore, let there be gratitude.

And so for this group of people, let it be our thought first to be grateful: we give thanks for the earth, Our Mother, as we are related to it, that which supports our feet, and so it will be in our minds.

Bodies of Water

And now we will speak about what is contained by the earth. He decided, "I will leave everything there." And he also decided, "There will be bodies of water here and there, large bodies of water. And, too, there will be rivers, flowing streams and lakes here and there. And there will be springs also, and the earth will be strengthened by them. And also we will be obtaining from the springs that which refreshes us, the people living on earth. This is where people will obtain their happiness."

And we believe it is so still, For even this he planned, the bodies of water. And we continue to get from the earth the water we use.

There are so many uses that we get from it. Let it be our thought that we do it carefully, for our happiness comes from there. He decided, "There will be bodies of water here and there on the earth."

And now carefully we thank him, he who in the sky dwells, Our Creator, and so it will be in our minds.

Grasses

And then he decided, "I will plant many kinds of grasses that will grow there. They will all be of different heights, and the manner of their growth will be different. They will appear in different array. And when the wind turns warm, people will see new growing things, many kinds of growing things, that will give them pleasure, those living on the earth. Even including children.

"And the breeze (i.e., the breeze carrying the scent of flowering things) will cause people's happiness." And it is so still, for when the wind turned warm, we saw them all there, the grasses growing that he had planted on the earth.

He decided, "This will be something important." For indeed he saw the distress from sickness, of the people moving about on the earth. And so carefully he assigned something to them, the grasses he had planted. He decided, "Medicines I will make from them." And it has come to pass that he has made them. He decided, "All of them will have names that the people living on the earth will know them by. And also there is a certain way they will assist people. If it should happen in the families that there is distress from sickness, people will be able to draw on them, to use them. And it will be possible for people to be happy many more days, for them to continue to move about on the earth." And surely he is still sending the planted things. He decided, "Medicine I should make." Also, we are still obtaining our happiness from them.

So let it be our thought to be grateful that he is still sending them, the grasses that grow on the earth, and the medicines he made, so let it be our thought to thank him, and so it will be in our minds.

Hanging Fruit

And now we will speak again about what he himself has done, he who in the sky dwells, Our Creator. He decided, "There will be fruit hanging where the grasses grow on the earth, beginning just above the earth. When the wind turns warm, people will see new fruit hanging again." It is the first thing they will see, how they come in succession, the hanging fruit he left on the earth. He decided, "It will be important to them, the people living on the earth. And they will be gathered to thank me when they swallow again what is new (referring to the Strawberry Festival)."

And this also will be important. He decided, "The medicines I make will make possible their happiness, the people moving about on the earth." And certainly he is still sending them. For when the wind turned warm, we saw now things growing again. And it came to pass that the fruit were hanging again. And also it was possible for us to be gathered to swallow again the hanging fruit he gave us, and it was possible also to thank him. And that is how it begins, with what he himself has done. He decided, "They will grade up-wards in sequence. (I.e., different kinds of fruit-bearing plants appear at different times during the course of the summer. First come the berries "just above the earth," and these are followed later by fruit-bearing bushes and trees.) People will go on seeing the fruit hanging: I have left it all for them, the people living on the earth." And it is so still, for during the last season we saw them all, how they came in succession, the hanging fruit he left, and we were able to swallow them again.

He decided, "It will be important to them. People's happiness will be coming from them." And we think that it has all come to pass as he planned it. Surely he is sending them in succession, the hanging fruit that grow on the earth. And let it be our thought, that carefully we thank him, he who in the sky dwells, and so it will be in our minds.

Trees

And now we will speak again. He decided, "There will be trees growing here and there on the earth, and also there will be forests of trees standing in array, and groves standing on the earth." And indeed there are trees still growing here and there, and the trees are standing in array.

He decided also, "This will be something important to people, for they can become medicines." And surely they are still growing, though the manner of growth is different for each one. He decided also, "All of them will have names, every one, that they will know them by, the people living on the earth. Also it will be possible for people to obtain their happiness from them in the families. People will depend upon them when the wind changes. For when the wind turns colder people will be able to keep warm." And they will work together as one, the live coals he left on the earth. He decided, "They will work together to bring happiness to the families on the earth." And we think that it is still coming to pass with regard to this matter.

And carefully he also decided, "They will have a leader. People living on the earth will say, 'It is there, the maple, it is a special tree.'" He decided, "When the wind turns warm, the sap will flow and then the trees will be tapped and the sap will be collected and boiled down by the people. It will be possible then for people to swallow it again. And it will be possible for people to be gathered." And it is important also. The medicines he made will be useful to them, the people moving about on the earth.

Now when the wind turned warm, we did indeed see again the new sap flowing, and it came to pass that we did swallow it again, and it was possible for us to be gathered at what has been called the Maple Festival. And also we did thank him in the manner he prescribed we should always thank him at ceremonies. And we think the ceremony did go through.

Let it be our thought that we will be grateful, for surely he is sending them, the trees standing in array on the earth. He meant them to be for our use, those of us moving about on the earth. And carefully now we thank him, he who in the sky dwells, Our Creator, and so it will be in our minds.

Small Animals

And this he left also. He decided, "I will leave wild animals for the people living on the earth. It will begin with the small animals which will be of different sizes among themselves, and which will be of many different kinds. And all of them will have names, by which they will know them, the people living on the earth. And among them there will be one to lead them. People will say 'deer.'"

And there are still glimpses of them running about, the wild animals on the earth. Surely he is sending them. He decided also, "There will be certain men—warriors—moving about on the earth. It will be their responsibility to keep themselves amused during the day, those moving about. And they will knock them down. And moreover it will be possible that people in the families will be refreshed, and that their bodies will be strengthened by it (animal meat), those moving about. And many uses will come to the people from there."

And we think truly it is going on as he left it. He decided, "The wild animals I leave will move about on the earth." And his promise extends to this: he decided, "There will be water on the earth, and there will be creatures moving about underwater, and these also are wild animals." (Including, perhaps, both water animals and fish.)

And up to the present time we are still seeing them; surely he is sending them; so let it be our thought carefully to thank him, and so it will be in our minds.

Birds

And this he left also. He is the one who did it, he who in the sky dwells, Our Creator. He decided, "There will be forests growing." He decided, "That is where I will leave them the many kinds of birds that fly about. And also there will be different sizes among them. It will begin with the small birds, and will go up to the large ones that can fly as high as the clouds." And still it is coming to pass up to the present time. For we see them flying about, the birds he left, the ones flying about in the forests.

And he decided also, "This will happen when the wind turns warm." And carefully he did it. "It will be possible for people to draw upon them for refreshment. And many uses will come to the people from there. And when the wind turns warm, the birds will be singing again, and it will make them happy, the people living on the earth. And the minds of the children running about will be strengthened. Everyone will be happy, the people moving about on the earth." And still he is sending them, the many kinds of birds that fly about over his world, so let it be our thoughts also . . .

And so he left them. He decided, "Many kinds of birds will fly about in the forests." And he is sending our happiness, our contentment, those of us moving about on the earth. Let it be our thought to do it carefully; surely he is sending them; now we thank him, he who in the sky dwells, Our Creator, and so it will be in our minds.

Our Sustenance

And now we will speak again. When he made people for himself, we who are moving about on the earth, He deliberated carefully on the matter: It would not be possible for them to manage alone, the people moving about on the earth. And so he left what are called Our Sustenance.

And for this he sent seeds. He decided, "This will sustain them." Moreover he decided, "There will be people in the families that will have hard work to do. It will be their occupation during the day, when the wind turns warm, to select garden spots and work the land. And then carefully they will place them underground, Our Sustenance. Then people will beg that it be possible that they grow well." (Referring to the Seed Planting Ceremony.) And indeed it has come to pass. For when the wind turned warm, people placed them underground in the gardens they had selected for planting. And then we saw them growing, Our Sustenance.

And people saw new things hanging. It was possible also for the people to be gathered. They have called it the Green Bean Festival. And we thanked him, he who in the sky dwells, Our Creator. Indeed, the ceremony was passed; we did thank him in the manner he prescribed we should always give thanks when people see anything he has given. And so it came to pass during our last season. And surely it went on, that they were growing, Our Sustenance, for indeed we saw them again. The Sisters, Our Sustenance, came in again. Then the people were able to be gathered at the great doings. He decided, "There will be a ceremony." And we think that it came to pass, that the ceremony was held during the last season. (Referring to the Green Corn Ceremony.)

It was a pleasure to see how far Our Sustenance had advanced. How lucky we were to see them grow to maturity again. And people were able to store them away. And truly everyone did give thanks. (Referring to the Harvest Festival.) So let there now be gratitude, for they still have their minds fixed strongly upon them (this antecedent reference is to the children mentioned previously), Our Sustenance.

There are, still, children standing upon the earth. And it is true, also, that they are still running about. And it is upon them that Our Sustenance have strongly fixed their minds. (The speaker commented: "The tyonhehkoh have their minds on the children because they are nearer to heaven. There will come a time when there won't be any children, and the foods will also go.") And we have been able to get all our well-being, our happiness, from them. This is why we are grateful. From Our Sustenance, the Sisters that he left, we have been getting many uses, day and night. They strengthen our bodies, we who move about on his world. So let us be grateful, let us think upon it, for surely he is sending them, Our Sustenance; and carefully we thank him, he who in the sky dwells, Our Creator, and so it will be in our minds.

The Thunderers

And that is how far it has been possible to go. We have spoken about what he created on the earth, how much he created that he left for our benefit, and our happiness: this we pointed out. And now we will speak again.

He appointed a series of helpers to have certain responsibilities. He decided that then no one would have too much to do. The first of these are the ones that come from the west called Our Grandfathers, as we are related to them, the Thunderers. He decided, "They will carry water." And at certain times we will hear them, the Thunderers. They are fulfilling their responsibilities. He decided, "They will continue to make fresh water, wherever there are bodies of water on the earth. Also, they will wash the earth now and then, and they will sprinkle what he has planted on the earth, that it may grow well. And people will be able to have gardens. It will come from them. And they will sprinkle water over them, and then they will grow, Our Sustenance."

And it is all still coming to pass, the extent to which they have been carrying out their responsibilities. So let it be our thought that there be gratitude.

He decided, also, that we would be moving about over this world. And he gave them the authority to keep below the earth all of the ones he himself did not create, for it has been proven to us that if they could move about on the earth, anyone even looking at them, the monsters, could be swept away from the earth.

And they have been carrying out their responsibilities; so let us therefore be grateful; let it be our thought carefully to thank them, from the west they come, the Thunderers, Our Grandfathers as we are related to them, and so it will be in our minds.

The Sun

And this is what he did. He decided, "There will be two periods on the earth: A period of light and a period of shadows. And when it is light, during the day, he will be in place in the sky. We call it the Sun, Our Elder Brother, as we are related to him." It is he who makes it light. He keeps moving about us, and it is true also that he gives warmth to the land. He keeps the wind at a certain velocity for the happiness of those of us who are moving about.

And so it is still. He is doing everything he was assigned to do. We can see that the light is bright when he is in place in the sky. And we are happy, those of us moving about on the earth. So let there be gratitude, and let it be our thought that we do it carefully. Let us be grateful that he is still doing his duty, that he is able to keep growing what the Creator planted on the earth.

And also Our Sustenance have been growing well under his care, and they are growing well, Our Sustenance, where he has been warming the earth. So let it be our thought that carefully we thank him, the sun, Our Elder Brother, as we are related to him, and so it will be in our minds.

The Moon and Stars

And indeed we have been hearing of the two periods he established. There will be a period of shadows every day. It will get dark. It will be just so long between the times that she is in place. He decided upon her tasks when he designated her the nighttime one, the Moon.

He decided, "The Moon will have phases. She will keep moving about us. People will determine time periods by her, the people moving about; moreover, people will determine the arrival of children on the earth. She will be helping to raise them also, and she will control the dew which falls on the earth at night."

That which he planted on the earth will be growing. She will be looking after Our Sustenance by letting the dew fall on them. She is carrying out the responsibility assigned to her. So let there be gratitude.

They are working together, the stars arrayed next to her, the Stars in the sky. In days gone by men used to know their meanings. But today, up to the present time, we do not know what their responsibilities are, the true responsibilities of the Stars next to her. Yet it is still satisfying to see them working together in the sky next to her, the nighttime one, the Moon, the Stars in array; into one we group them, they are still carrying out their responsibilities; and carefully we thank them, and so it will be in our minds.

The Four Beings

And now we will speak about this. He is the one who did it, he who in the sky dwells, Our Creator. He decided, "I will be assisted by the Four Beings, 'his helpers.'" We call them the Sky Dwellers. They have certain responsibilities. For indeed at one time he saw that the people were of many minds, the people moving about on the earth. And so he spoke, telling them all that was on his mind. And they have been carrying out their responsibilities up to the present time. He decided that they would guide us, look after us. For it is possible that sometimes people are not happy. But they have always been able to guide us, and they have indeed been carrying out the duties he assigned them. He decided what would be their responsibilities. It is possible also for them to straighten out the mind each of us possesses. Indeed, it has been proven to us that they are hovering just above our heads. And our happiness has been made possible. So let it be our thought, that carefully we thank them, the Four Beings, his helpers, and so it will be in our minds. And our happiness has been made possible. So let it be our thought, that carefully we thank them, the Four Beings, his helpers, and so it will be in our minds.

The Wind

And now we will speak about this. He is the one who did it, he who in the sky dwells. He decided, "There will be winds moving about on the earth that will have a certain velocity for the happiness of the people moving about on the earth. There is a certain place where the wind originates, a 'veil' people will call it. That is where the wind originates. And it will have just the right velocity for people's happiness."

And it is still coming to pass that up to the present time we are happy. And now carefully—for it is his own creation it will be our thought to thank him, he who in the sky dwells, Our Creator, and so it will be in our minds.

Handsome Lake

And now we will mention this. He used to stay here in those days. (Handsome Lake is not actually named until the end of the section.) He used to move about on the earth. It was then that he saw, he who in the sky dwells, Our Creator, what was present on the earth, that it was possible for people to stumble, the people moving about on the earth, when they should have seen the basis of their happiness. So carefully he brought his word down.

He decided, "This will be solution for people in the future." And so he picked him to carry it, the word of Our Creator. He fulfilled everything expected of him. Indeed he planted the message among the people living here and there. He moved about among them, laboring for many days telling the word of Our Creator.

And up to the present time we still hear it, there is still the sound. And surely it is at hand, it extends to us here. And when he spoke (i.e., for the last time before his death), this is what he said: That he had done what he could, that it would come about as he had spoken of it. And this he said, "It will be known by everyone where I will be returning to. Let it be thought that everyone should follow in my footsteps."

And it has come down to us, the Good Message, the Creator's word that he carried around. He fulfilled what he was able. It has been proven that he is happy again in the Creator's world. And now carefully we thank him, our great one, Handsome Lake, and so it will be in our minds.

The Creator

And now we will speak again, about him, he who in the sky dwells, Our Creator. He decided, "Above the world I have created will be the ever-living world. And I will continue to look intently and to listen intently to the earth, when people direct their voices at me."

And now the speech, that you heard come straight out as far as it was possible, is becoming difficult. Truly he has been listening intently to us. And if it happened that we left something out, it would never be diminished for him who in the sky dwells, Our Creator. For it is right in our minds. Let there be gratitude day and night for the happiness he has given us. He pities us also, and he loves us, he who the sky dwells. He gave us the means to set right that which divides us. And we may still have our happiness. Let this alone be in our minds. And it will be our thought, let us be grateful that so many people are happy, and into one we carefully put our minds. And now we thank him, he who in the sky dwells, Our Creator, and so it will be in our minds.

Epilogue

And that is how far it was possible for me to bring it out. We directed our voices toward him, he who in the sky dwells, Our Creator. And that is the best that could be done. Let it be our thought that we will abide by his word alone, for if we do, day and

night, we may yet be happy. And that is how far it was possible for
me to carry out the responsibility of this speech. So be it.

Enos Williams
(teHanraHtihsokwa, "Quivering Leaves")

A mature speaker of the Turtle Clan (Wolf moiety) at Seneca
Longhouse (Six Nations). Williams's tribal background is Mohawk;
his father was Cayuga. He spent his early years with his maternal
grandmother who spoke Mohawk with him. That was his first language. He learned Cayuga later, and Onondaga from his wife later
still. He now prefers to use Cayuga in the longhouse, but uses the
other languages on occasion. . . . He lives a considerable distance
from the longhouse, about two miles. He is now the principal speaker
for the Wolf moiety and is "head faithkeeper" (hotrihotko:wah) for
that side. . . . He reports having started speaking on informal occasion such as work bees and longhouse socials at about age 17.
(Foster: 35–36)

Appendix 3

Ernest Smith Paintings—Rochester Museum, 1991

The author has placed the 240 paintings by Ernest Smith into categories: the Thanksgiving Address, Creation Story, the Great Law of Peace, the Code of Handsome Lake, Historical Interaction, Ceremonial, Stories, Persons, and Miscellaneous. Descriptions of the paintings have been provided by the author, and where available, Ernest Smith and/or Arthur Parker descriptions of the paintings have been quoted.

Please note that the author has not provided any descriptions of the False Face and Ceremonial paintings out of respect for our peoples' belief that these topics are not to be shared with the public. In the past, when we have shared this information, it has been misconstrued in a disrespectful manner by researchers who did not understand our culture. Therefore, we no longer discuss private healing matters with the public.

BC indicates buckskin clothing indicative of pre-contact era.

CC designates cloth clothing of the post-contact era.

BL means bark longhouse both curved roof or peaked roof.

Thanksgiving Address

1. The People

545   Sky Woman—The Creation of this world.

1108 Lunchtime—The Mother or Grandmother is calling two children to eat, and they run toward the BL as if racing. On the bench are corn bread, corn soup, and a pottery vessel is on the fire. She uses a wooden ladle. BC, pottery.

1116 Children At Play—A boy and girl have constructed a small stockade with village inside, moat surrounds village, corn husk dolls, CC.

1119 Adoption of Orphans—Family (mother, father, young boy and girl) leaving the site of two graves. The little girl is weeping. A boy and girl run from the village toward them to greet them. The woman is pointing at the two children as if to say, Here's your new family. CC, BL (both curved and peaked roofs). "There is always a family to take care of an orphaned child. The moral is there is unity among the people" (Ernest Smith). (Images from the Longhouse, brochure)

1050 The Storyteller—Older man sitting under a tree talking to a teen-aged boy who listens intently. BC.

736 Mother with Baby—Woman, child, CC, cradleboard.

1112 Clean Camps—Women sweeping out the bark longhouse, CC, in the background a woman is using a wooden ladle to scrape something from a bark tray into a refuse pit, very neat, orderly village.

2. Mother Earth

75.224.1 Conservation—Father and son each have two fish as they leave the stream, father holding up two fingers, the son seems to be pointing back at the stream as if to say there's a lot more fish, but the father is saying two fish each is all we need, BC.

1102 Child Studying Nature—Young boy sits quietly observing a raccoon fishing in a stream, farther up the stream two deer cross the stream, there's a squirrel in the tree above him, and a bird in a nest.

685 The Potter—Clay from the earth, BC, woman makes pottery, a man tends the fire where the pots are being fired, he shields face as if to protect himself from the intense heat, BL.

691   Two Men Dipping Bowls—Sequal to 685, the men have removed the pottery from the fire with sticks and are putting the pottery into bark tray full of water to cool bowls.

717   Making Pottery—Mother and daughter making pottery, BL, BC.

3. Bodies of Water

704   Dipping Water—One man, BC, at a small waterfalls dips container with the flow of the water, this is more landscape.

741   Two Women Carrying Water—Smiling women dipping water with pottery containers, CC.

681   Burning Springs—Natural hot spring, one man, BC, lights fire on the water.

1131   The Fisherman—Two men, BC, fishing.

4. Grasses/Plants

1022   Making Sunflower Oil—Two women, corn pounder, grinding stone, pottery, BC, bark basket.

1002   Gathering Artichokes—Two women, splint and bark baskets, BC, digging up artichokes.

719   Medicine Picking—Man, CC, and woman, BC, in the woods, basket, burning tobacco.

1003   Gathering Oyster Mushrooms—Two women, old log with mushrooms growing on it, drying rack, BC, baskets.

5. Hanging Fruit

1688   Jungie Turning Strawberries—Two little people, BC, (Fantasia style.)

1010   Picking Wild Strawberries—Happy smiling family (mother, father, child), birchbark baskets, BC.

1019   Picking Black Raspberries—Mother and child, CC, baskets.

6. Trees

709   Maple Sugar Time—Man, woman, BC, wooden spiles in tree and sap drips into bark and wooden baskets, he collects sap in pottery, woman has three large pottery containers of sap boiling over fire, four large flat rocks are standing behind the fire and reflect the light and heat, snow on the ground.

653   The Spirit of the Wind Meeting the Spirit of the Maple— Wind is male, he swoops out from behind tree, maple is female, she seems to step out of the tree trunk, BC.

716   Giving Thanks to the Great Spirit for the Maple Sugar— Ceremony outside longhouse, many people, CC, pottery.

1133   Maple Sugar Implements—Tools for making maple sugar, wooden spiles, bark and wooden trays, pottery.

680   The Spirit of the Pine Tree—Male spirit in one tree, one woman, older, gray hair, looking up at spirit, BC.

655   The Frost Spirit—Spirit appears as man made of ice, taps tree with war club.

1130   Making Hemlock Tea—One man, BC, putting hemlock tree branch into potttery over fire.

1020   Gathering Chestnuts—Family (mother, father, child), mother is putting chestnuts in splint basket, child grinds chestnuts on grinding stone, dog looks up in tree as father throws wooden club into tree to knock down chestnuts.

1012   Drying Plums—Baskets, two women, CC.

1021   Making Hickory Nut Oil—Mother, son, pottery, grinding stone, BC.

683   Longhouse Under Construction—Three men, BC, building BL.

711   The Basket Maker—CC, one woman, splints, basket.

7. Animals

743   Story of Red Hand—He (a young man) is fixing a tree with a splint, deer, raccoon, otter watch as he repairs tree.

734   Story of Red Hand—Snarling bobcat has just killed a rabbit, Red Hand, BC, holds up one hand and points to rabbit

715   The Little Water Society—Many animals make medicine for the wounded Red Hand, BC.

1078   Thanking the Spirit of the Bear—BC, man has built a fire, breaks an arrow over the fire, a bear he has killed in the background.

732   Bear Dance—Man covered with a bear hide which seems to be tied, and he leans over to eat from basket of berries on a bench in the longhouse, dancers men and women, CC, one singer.

737   The Chipmunk—Wooden bowl, ladle, corn soup, chipmunk eating from bowl, the lesson is sharing.

1066   Animals of Prey, The Hunter and the Hunted—Man with bow and arrow pointed at the bobcat, who is in a tree ready to leap onto a fox, who looking at a rabbit who escaped, winter.

1049   The Hunter—Man, BC, bow and arrow, deer.

950   The Story of the Hunters—Father and son, BC, hunting wild turkeys.

1001   Preparing Smoked Vension—BC, man cuts up venison, woman puts strips of venison on drying racks.

1088   The Wounded Bear—Story of big dipper, three hunters and bear.

1692   The Nephew and His Cruel Uncle—Story, lots of animals, BL, uncle seems to be sending the boy into the woods alone, BC.

955   The Hurt Warrior—A bear is tenderly holding the arm of a hurt warrior, an eagle looks on, BC.

1052   Fighting a Bear—BC, man with stone hatchet versus a bear.

8. Birds

714   How the Birds Got Their Color—Bird is sitting on wolf's nose, birds in trees, racoon leaving.

1064   The Vulture Chooses His Feathers—"He (the vulture) had the choice of feathers, but he was in such a hurry to fit himself with nice plumage that he put the outfit on and it's too short for him. . . . He was in such a hurry to pick out a cloak that he picked the wrong one. . . . I think the moral is, 'don't be too fast to pick out something that may not suit you' " (Ernest Smith). (Iroquois Trail brochure)

619   Dance of the Eagle—Eagle has talons on dead deer, man singing, BC, did man offer deer to eagle?

1107   The Pigeons—Man chopping down tree full of pigeons, woman in background putting meat on drying rack, lean-to shelter, CC.

1129   The Nest Robber—Boy in tree stealing robin eggs, his mother is scolding him, she has her left hand on her hip and with the right hand is pointing down, CC.

1000   The Trapped Eagle—Eagle is in a pit, one man, BC, reaching into the pit with a small loop on a pole.

ACC 75.225.1 Hunter Waving to Geese—BC.

9. Our Sustenance, Three Sisters— Corn, Beans, Squash

545   Sky Woman—Village in Sky World, Sky Woman falling, Gahaciendietha , fire beast, gives her corn and corn pounder, birds coming up to catch her, turtle, water.

721   The Three Sisters—Three female spirits in corn, beans, and squash, little person turning the squash to ripen.

1014   The Three Sisters—Same as above without little person.

610   The Bread Dance or Women's Dance—Moon Dance, first woman carrries a little turtle rattle, the second woman carries corn, more women form line of dancers, in the center is iron kettle of corn soup, CC.

511   Blessing Cornfield—BL with one smoke hole indicating an individual family dwelling, corn field with hills, virgin naked, basket, pottery, man, CC, singing at edge of field, another woman, CC, in background.

1084   Introduction of New Corn—CC, man with bundle of ears of corn on his back is handing ears of corn to a smiling family consisting of mother, father, girl child, grandmother.

497   New Year's Announcers—Bigheads carry pestle, corn husk braids around waist and ankles, log cabin, two tracks like road.

712   BigHeads—Two men in Buffalo robes with corn husk braids around waist and ankles, carry corn pestles.

637   The Husk Dancers, The Dance of the Husk Mask—Masks are made of corn husks, a braid of corn hangs on wall to the left, two dancers, breech cloth, lean on tall thin sticks, one man, CC, holding up a wooden spoon.

494   Eagle Dance—Dancers, corn pounder in corner, CC and BC.

690   The Corn Spirit—Cornfield, one man, Handsome Lake, female corn spirit drapes corn leaves over his shoulder, his hand reaches up to touch the corn leaves, CC.

——   Hoeing Corn—Three women, CC- fancy beaded, hoe is stone or bone tied on stick with sinew.

744   Corn Gatherers—Two women, corn in bundles on their backs, CC, walking home from cornfield.

516   Story of the Red Ear—Two women braiding corn, sitting in corn field, they are laughing as the one points to ear of red corn the second woman has husked, piles of braided corn on the ground, harvest, baskets.

1100   Working Corn—Woman, CC, scraping corn off cob with deer jaw bone, bark basket.

740   Washing Corn—Woman, CC, corn basket, pottery. "The white corn of the Iroquois was used for their hominy, their bread and their gruels. The hard, ripened kernels were boiled in a wood-ash solution until the hulls became loosened, when they were washed off in a special basket. To the Iroquois maize was a sacred plant which must be propitiated by ceremony and treated with respect" (Parker 1939).

1104   Roasting Corn—Three bark baskets full of corn leaning on rocks, next to fire, to dry the roasted corn.

1127   Denial at the Storehouse—Man approaches the grainery with his hand out for corn, another man holds up his hand to stop him, another man walks away with braid of corn over his shoulder, the grainery is round bark structure with peaked roof on poles with hand hewn log ladder, all BC.

1023   The Boy Who Popped Too Much Corn—Boy running away from longhouse exploding with popcorn.

1065   Praying for Manna—(from heaven)—It's night and three people sit outside the BL with baskets. Corn kernels and ears of corn are falling from sky.

1685   Corn Maiden Is Burned When Corn Is Burned—A man is throwing an ear of corn and corn bread into fire, a woman is holding out hand to say, no, don't do that. (Tuscarora corn story)

10. The Thunderers, Our Grandfather

724   Heno, The Thunder God—Heno has bow and arrow, he is a spirit in the clouds. "Heno, the thunder god was the enemy of evil spirits and of monsters for whom he had a special dislike. His flint-tipped arrows made the lightning, though at times it leaped from his eyes or throat; his voice was the thunder's roar" (Parker 1939).

633   The Hail Spirit—Spirit throwing hail from pottery, man CC covering his ears. "The Hail spirit cannot be seen but he hurls his frozen pellets with devastating force. Hail storms do not last long because Hail Spirit's pot soon becomes empty" (Parker 1939).

656   Thunder God and the Spirit—Thunder God in sky sends lightning from his mouth toward horned serpent in water.

1077   Searching for the Rainbow—Two men, BC, one chasing the other toward the rainbow, stream in background, sun shining behind rainbow.

11. The Elder Brother Sun

Going to Visit the Sun Father (See Parker 1923:65)

1067   Braving the Sea

1068   Swimming the Cataract

1069   Removing the Boulders

1070   Going Through Fire

1071   Blown by the Wind

1072   The Meeting

Drying foods: 1011 venison drying, BC; 1012 plums drying, CC.

12. Grandmother Moon and Stars

651   The Elk That Fell in Love with the Morning Star—Woman in sky, bright light behind her, elk gazing up at her.

1015   The Seven Dancing Stars—The seven boys sing and dance as they rise up into the sky and become Pleiades.

1088   The Wounded Bear—Three hunters, bear, moon in sky, story of the origin of the Big Dipper.

651   The Elk That Fell in Love with the Morning Star

1109   Woman in Seclusion—Village, separate wigwam in which one woman sits.

13. Four Beings—Not Illustrated

14. The Wind

653   The Spirit of the Wind Meeting the Spirit of the Maple— Wind is male and swoops out from behind a tree, Maple is female who comes from the maple tree.

742   The Daughter of Sky Woman—Who gets pregnant by the West Wind.

15. Handsome Lake

630/631   Handsome Lake and the Three Messengers—Plank board house, Handsome Lake leans on doorway in which appears three spirit like beings.

16. The Creator—Not Illustrated

Creation Story

545   Sky Woman—Shows village in the Sky World, she's falling, fire beast gives her corn and corn pounder, birds, animals, and the turtle, BC.

1011   Sky Woman—Shows only one longhouse in the Sky World, omits the fire beast.

742   The Daughter of Sky Woman—She gets pregnant by the West Wind shown as spirit, she's naked.

1092   The Twins—Sunbeam shines on one twin who walks up the sunbeam, corn pounder, woman cooking soup in pottery, BL, BC.

618   The Great Feather Dance—CC, dancers, turtle rattles,
          wooden bench, inside longhouse.

696   Lacrosse Game—Shows many players, BC.

Great Law of Peace

613   Hiawatha's Sorrow—Daughter's body laid out under tree,
          plus six more daughters' graves in background, BC.

546   Peace Queen—BL, BC, corn decoration on breechcloth,
          looks like two groups of people meeting, possibly Eries
          (panther skin headdress) and Haudenosaunee (Mohawk-
          style haircuts), she stands at center of two groups, man
          near her drops his war club.

616   The Council with Tadodaho—Snakes in his hair, war club
          in his hand, Peacemaker and Hiawatha, wampum belt
          and strings of wampum on ground, flint tipped arrow, BC.

681   The Wampum Maker—One man, BC making the
          Hiawatha belt.

723   Totadaho, Hiawatha, and the Peacemaker—Wampum belt
          has heart in the middle.

Code of Handsome Lake

630/631   Handsome Lake and the Three Messengers—Handsome
          Lake skinny man holding onto the door frame of plank
          board house, sees three spirit-like persons.

524   The Vision—Spirit-like longhouse village in trees, can see
          people in this spirit-like village, two men, one sitting under
          a tree, the other is pointing to this spirit-like village, BC.

707   Handsome Lake Preaching—Inside plank board long-
          house, CC, men on benches across men's end of longhouse,
          women on sides.

958   The Gluttons—Three fat men, one starving man. "Tem-
          perance in all things was an Iroquois principle. Gluttons
          were threatened with the appearance of Sadodawkwus,
          a lean ogre who spooned out their vitals and devoured
          them" (Parker 1939).

733 Spirit of the Gambler—Blanket, bench, peach stone game, man, a spirit-like being looks greedily at items left in this world.

1013 The Pathway of the Souls—BC, BL, four men by a fire, woman in the doorway, a hand reaching down from the sky, they all look at it.

1082 The Trail to the Great Beyond—One path full of rocks, the other path has a serpent, one old person stands where the two roads join looking down the one path, there is a person farther down the path, clouds, stars.

Historical Interaction

702 The War Party—BC, painted post with hatchets embedded.

1114 The Stockade—Plank over moat, man with bundle, trade with other Indians, BC.

1110 Trading Arrowheads—With other Indians for furs, displays variety.

1128 The Trappers—Two Indians, one in canoe, one on shore with pack of furs on his back, BC.

954 The Gift of the Panther Skin—Erie man wearing skin of panther as headdress, trading Indian to Indian, pelts, BC.

1022 The Captive with His First Pair of Seneca Moccasins— BC, war club, one man hands moccasins to captive who carries bundle on his back.

1117 The Dutch Trader—Piles of furs, and a chest filed with cloth.

Sullivan's Campaign—A Series of Four Paintings

698 Soldiers Shooting—At Seneca as they run, one dead, one on horse, one running, soldier chasing.

699 Scout Murphy—Frontiersman scalping a Seneca as soldier looks on.

700 Houses and Cornfields Burning—Soldiers burn two peaked roof log cabins, one Seneca man lying on ground dead, another wounded trying to get up, soldiers emerging from the cornfields.

701   Senecas Fleeing Burning Village—Soldiers riding in from the right, center the village and corn fields on fire, Senecas running away, women, children, an old man, one young warrior, a child grabs his hand and points back at the burning village.

583   Progress—Shows five Haudenosaunee men with horses, BC, on a cliff looking down on a white farm, steamboat chugs up river, train going over river on a trestle, off in distance smoke rises from large building with two tall smokestacks, only one row of trees in the valley.

ACC 76.13   Jim Hill's Place—Woman, BC, carrying water to log cabin.

Ceremonial

497   New Year's Announcers—BigHeads, log cabin, double tracks like a vehicle.

494   Eagle Dance—Inside plank longhouse, BC and CC, corn pounder in corner, splint basket, bark tray.

727   Eagle Dance—BC, outside, post with arrows, rattles, and bundle tied to it, one singer.

610   The Bread Dance or Woman's Dance—Moon dance, CC, woman leads with small turtle rattle, next woman carries ears of corn, others follow, iron kettle of steaming corn soup in center.

617   War Dance—BC, two singers, all Mohawk-style hair cut, weapons.

618   The Great Feather Dance—CC, bench, two turtle rattles, breech cloth, leggings.

637   Husk Dancers, The Dance of the Husk Mask—Two dancers lean on long skinny sticks, singer.

636   Sacrifice of the White Dog—BL, CC, many people, outside, tripod with white dog tied to it.

652   Peach Pit Game, Peach Stone—Men and women, CC.

738   Peach Pit Game—Women, CC, shaking peach stone bowl, kneeling on blanket, two men watch.

712   BigHeads—Two, buffalo robe, corn husk braid at waist and ankles, pestle.

716   Giving Thanks to the Great Spirit for the Maple Sugar—
        Outside, many people, BC, CC, pottery, fire. "In the old
        days this ceremony of burning tobacco in the fire in thank-
        fulness to the Creator was done every day" (Ernest Smith).
        (Images from the Longhouse).

1105   Preparing Tobacco—One man, CC, crushing dried tobacco
        and putting it in leather bags.

657   Idose—Block of wood with charm on it, gourd rattles, all
        men, BC, CC, singing/dancing, fire, at night, BL.

682   The Medicine Man—Individual ceremony, BC, CC, mixing
        herbs, BL, shows interior with the two levels of shelves.

706   Dawn Song—One person, BL, fire, smoke rises, BC, sings.

951   Purification Ceremony

1018   Dipping Medicine Water for Little Otter Society

1053   Dark Dance

1059   Quiver Dance

1680   Sweat Bath

False Faces

713   Dance—False Faces.

718   Inovocation to Our Grandfathers—False faces.

517   False Face Dance

571   The Carver—Of false face masks.

611   False Face Society

722   Healing Ceremony—False face, one beating stick on corn
        pounder, man burning tobacco.

725   Carving the Great Face—False face.

1103   Dreaming of the Mask Society

1091   The Thunderer

739   Lunette

689   Members of the False Face Society—More contemporary,
        has regular modern men's pants on.

Stories

1055   The Flying Canoe—Face in cliff.

1056   The Storyteller of the Cliff—Stone face in cliff.

1057   The Serpents

1058   The Tug of War

1063   The Stone Giantess—Female, has another woman's baby.

1073   The Man Who Conjures up Maidens

1074   The Toad Woman

1075   The Thin Man

1076   The Chiefs

1077   Searching for the Rainbow

1080   The Serpent

1085   Day Comes, Goblins Go—Fantasia

1082   The Trail to the Great Beyond—Fantasia

1086   The Running Moccasins—Fantasia

1089   Fantasia

1087   The Ghost Wife

1093   Old Crooked Mouth

1094   Hadodenon Finds His Elder Brother

1097   The Chipmunk's Stripes

1098   The War Party

1096   The Boycub

1106   The Mammoth—Woolly mamoth chasing people, BC.

1111   Witchcraft

1113   Boy Raiding Bee Tree—Old man, boy stung by bees.

1681   Legend of the Living Heart of Enemies

1683   The Magic Flight—Bear, birds, corn.

1686   Flying Light—Fantasia style.

1698   Sees His Reflection in the Water

Persons (Contempories of Ernest Smith)

581   Lyman Johnson

582   Benjamin Ground

684   Portrait of Sarah Hill—Iron kettle, CC, braiding burden strap.

687   Portrait of Harrison Ground—Carving wooden bowl, CC, gustoweh with silver band.

686   Portrait of Jesse Hill—carving war club, CC, BC, gustoweh with silver band.

688   Portrait of William Gordon—Carving False Face, CC, gustoweh with silver band.

Acc76.193 Jim Hills Place—Log cabin, woman carrying pail of water, BC.

Miscellaneous

Games

693   Hoop and Dart Game

694   The Snowsnake Game

729   Lacrosse Game—Fewer people.

1058   The Tug of War—Gustoweh with antlers, BC, two teams.

Making Useful Items

1061   The Bow Makers

1079   The Pipe Carver

1090   Gathering Wampum Shells

1099   The Burden Strap Maker

1124   Wampum Belt Making

1118   Bark Processing—Making splints for baskets.

1126   Making of Snowsnakes

1132   Flint Mining

Not Categorized

735   The Flute Player

1060   The Vision—Woman, two spirits fighting.

1095   Sketches of War Implements

1125   Silver Ornaments

Death

1121   Two Oneida Graves—Buried in mounds with stockade around graves and clan on stockade.

1135   The Wolf—Wolf is approaching, BL, a skelton?

703   The Night Watch—A Wake, BC.

720   Death Dance

1016   The Origin of the Dance of the Dead—One woman, stream, bat in tree.

1017   Sacrifice of a Maiden—Canoe going over Niagara Falls.

952   The Faceless Death Leads Out the Spirit of Man—Spirit holds out hand to dying man who takes his hand and his spirit rises, woman weeping.

Sketches of Types of Houses

1682   Circular Bark House
       Manhatten Bark House—Looks like longhouse.

1684   Abnaki House

1689   Saux & Fox Lodge

1690   Ojibway, Menomini House

1691   Cree, Winnebago House

1692   Abnaki Headdress and Costume

1693   Navaho Hogan, Pawnee Lodge

1695   Wichita Houses

1696   Saux & Fox Lodge, Menomini

Appendix 4

Publications Containing Ernest Smith Paintings

Barreiro, Jose. *Indian Corn of the Americas, Gift to the World.* Ithaca, New York: Northeast Indian Quarterly, 1989.

Beauchamp, William M. *Civil, Religious, and Mourning Councils and Ceremonies of Adoption of the New York Indians.* New York State Museum, Bulletin 113, June 1907. Rpt. 1975 (cover).

Canadian Indian Marketing Services, 145 Spruce St., Ottawa, Canada (Reproductions).

Genesee Valley Council on the Arts. *A Genesee Harvest, A Scene in Time, 1779.* Rochester, New York: Early Arts in the Genesee Valley Series, No. 7, 1979.

Hauptman, Laurence, M. "The Iroquois School of Art: Arthur C. Parker and the Seneca Arts Project, 1935-1941" *New York History* (July 1979): 282–312.

————. *The Iroquois and the New Deal.* Syracuse, New York: Syracuse University Press, 1981.

Hayes, Charles, F. *The Iroquois in the American Revolution.* 1976 Conference Proceedings, Rochester, New York: Rochester Museum and Science Center, Research Records No. 14, 1981.

Judkins, Russell A. *Iroquois Studies: A Guide to Documentary and Ethnographic Resources from Western New York and the Genesee Valley.* Geneseo, New York: State University of New York at Geneseo, 1987.

Seneca-Iroquois National Museum. Allegany Indian Reservation, Salamanca, New York, brochure.

"The Iroquois," *Faces,* Cobblestone Publishing, Peterborough, NH. Vol. VII, no. 1 (September, 1990).

Underhill, Ruth. *Red Man's Religion.* Chicago: University of Chicago Press, 1965.

Appendix 5

Interview with Irv Powless, Onondaga Chief,
Onondaga Nation, New York, by Carol Cornelius,
November 27, 1991

Carol: I came to ask you about corn.

Irv: Corn, probably, is most important to our people because it's used in the ceremonies and then for enjoyment also. Corn is such an important factor. To use corn as a teaching. . . . You have a combination of three teachers. One, you have your culture teacher explain about corn, what it is, how it grows, the stories that go along with the corn. In the storytelling you'll draw a vision of the story. So now you need an art teacher. The art teacher will then take the story and put it on to . . . to show how the story evolves whether it be a series of pictures or whatever, and then your language teacher comes in and puts in all the language that's necessary for that story. The names of the corn, you got your white corn, yellow corn, and colored corn. You have the purposes for each one of these. You have the names. You have the corn basket. You have the wood ashes. So all of these things have names and so each one gets their Mohawk, Oneida, or Onondaga title written out and so you learn the language. You learn how to make corn soup, roast corn—each one has its name, so you have a whole list of language words that support the story as its told in English. So now you got language, art, and then you have math.

245

Because an ear of corn comes from a kernel of corn. You plant one kernel of corn and, if you're lucky, one stalk will grow. On the stalk will be two ears of corn. The two ears of corn will be thirteen inches in length with thirteen rows around, with how many kernels in each row. So from one kernel, multiply the dimensions, how many kernels of corn do you get from one? Now, if you're planting in a group—look how quickly it multiplies—and then—the financial—if you're into selling corn as a commercial thing—look at what your profit is. So, mathematically you'll be able to figure it out. You'll be able to go into an economic development program and show how this all works.

You have your art as an expression of yourself, also that comes into an economic way, and also how you can market those things. And then the cultural part and the spiritual, is the most important for the society. The feeling that corn is a very necessary element of your society and should be respected and so you have ceremonies for corn.

You have songs and dances. And then you explain the usage of the corn and then you have, as I said, the language. So, you can expand. And then, for drama you put the story into a play. There's various mechanisms which you can use to do that, . . . people, make puppets, make a stage.

The whole thing, starting with an ear of corn branches out into a vast education project. And it can be used not only with our people. Parts can be used with our people and parts can be used with the non-Native society.

Carol: What do you see as the difference between teaching our own children and teaching non-Native children? If you had five Onondaga kids, who said teach me what is to be Onondaga, and if you had five kids from the public school—

Irv: With the non-Native society you generalize, you teach historically. But when you're teaching your people, then the spiritual goes along with the historical.

Carol: I always worry about that because the cultural part is left out, then kids are stringing macaroni beads and making paper bag headbands. And, they are left not understanding us, and reduce it to macaroni beads.

Irv: You can do that same process with any one of the things that we have in our schools. You can teach about making a basket. So you bring in a basket maker to explain about making a basket. The language teacher teaches the name of the tree, the ceremony that goes with taking of the tree and then the splitters, what that's called. The basket after it's finished—because we have different kinds of baskets, and the tools what they're all about. And then you get your art teacher to draw the story of basket making and you illustrate usage of the baskets. It's all culture-based.

Carol: Culture-based curriculum?

Irv: Yes, it's all culture-based curriculum, see. Now you can do this with a water drum, you can do this with rattles, with clothes, with everything, see, because every time you bring in someone to make something, all of the language goes with it. Gustoweh, the feather, the hat, the makings of the hat, the birds that the feathers come from, the hawk, the eagle.

So you take one little item and from each of these items you can encompass the making of something, the ceremonies that go with it, the language that explains all of this, and then art to create a visual. And then if you want to do a play on it, you got the drama and then then you have story writing which gets to the English part.

You can do this with each one of them, write a story about it and then correct the English language.

Carol: If we were teaching this with non-Natives then, we wouldn't go as in depth into the spiritual?

Irv: That's right, because they wouldn't understand it and it's not necessary.

Carol: Would you teach them about the Thanksgiving Address?

Irv: Yes. In the talk that I did, the isolated part of it, showed the depth the Thanksgiving Address goes into when it's done correctly. What I presented at Cornell the other day, took about the same amount of time as the Thanksgiving Address when it's given. And I only covered one section in depth. So when they sit down and do one of these things it takes a hour and a half. Now you're getting into the protocol. What's usually done in these short versions that goes, "now we notice that the trees are there and they are still doing their job and we give attention to the chief of the trees the maple tree. It gives us sugar and syrup. So let us put our minds together and give thanks to the trees." It's done. But each one of these we can expand, this is what we do. Even my father, at the longhouse . . . the opening, and when he got through, my father says, "all of the years I've been sitting here listening to the opening—the man that just presented, this was the best that I ever heard. It was so nice the way he explained it." So everyone has their way of doing it, each one has their own expression, so it's different every time it's presented.

Carol: Michael Foster recorded Enos Williams.

Irv: He was a good speaker and a good singer. He gave a real nice presentation.

Carol: Does it make any difference, the order in which its presented?

Irv: It usually begins with the people and Mother Earth.

Carol: The Three Sisters seem to come in at different places, does that make a difference?

Irv: Not really, you know it's all individual preference in how they do their presentation, but usually they start with the people and Mother Earth.

Carol: When you were growing up did the people eat corn soup more regular for daily meals, not like now where it's eaten mainly for special occasions?

Irv: I think it's about the same today as it was a long time ago because there's so many ceremonies. So, you're always getting that. I think there's more awareness that this is a very tasty food that can be eaten in between ceremonies.

And so, for economic reasons, when I was growing up, we had people who would make white corn soup, white corn bread, and went around the community and sold it. So we were always getting corn soup in between the ceremonies.

We still have those people in our community who have a lot of corn. They'll make up corn soup and they'll share their gift of corn making because some of our people don't know how to make corn soup. Or, they make corn soup but it doesn't taste the same as someone else's corn soup. (laughter) You have corn soup makers, and *then* you have corn soup makers! (laughter)

Carol: Most people use wood ashes to clean the corn, but some use lye. Is there a difference?

Irv: There's a difference. The wood ash corn soup makers, when they make corn soup there's a knack—which comes with everything. So with these things that are there we have people in the community who make them, corn bread and sell it. It always sells. So we have that in addition to our ceremonies.

It's about the same, growing up as it is today. It's still there. If you're fortunate enough that your wife is one of those who makes corn soup, then you have it more often. It's a lot of work and you need to have water.

Carol: We used to make corn soup out at Cattaraugus and sell it at the local store and it always sold.

Irv: It always sells. On Sunday afternoon, we have corn soup makers who put a sign on their tree, "corn soup today," so we go there.

Carol: Do you see a pattern of raising corn that it stays in certain families from generation to generation?

Irv: Yes, it stays in the families, mostly your farmers. My grandfather was a farmer, he had a team of horses so he taught my father how to plant. And then my father had me out in the garden pulling weeds, I complained, but in the process I learned how to plant. And I plant a garden now.

Then you have some who don't plant anything. Who don't have the joy of being able to walk over to the garden and pick peas or carrots or other vegetables.

I turned my soil the other day. Before I done it, I'd dug up the carrots. What I do is, I broadcast carrot seed and I don't weed it. I planted five packages of carrots and I don't weed it. So when I went out there to harvest, I have big carrots and small carrots. So, anyway, I turned the soil and there were a lot of small carrots. My wife said to me, we need some carrots. And I laughed and said they're just laying out there. So she went out there and came in with a handful of carrots. And those little ones taste so good.

But you see, if you don't plant you miss that. If you're not into planting, you buy can goods, you don't buy the fresh vegetables.

Carol: There's a connection there—to the earth—

Irv: Yes, and to the Creator. If you're raised on can goods that's what you eat, and if you're raised that way you don't plant. If you eat canned goods you get all that salt into your system. Canned goods are not healthy.

With the harvesting of Mother Earth's gifts, right out there the very first are the leeks, then wild onions, cowslips, milkweed, and then berries. You know you watch as these gifts appear.

Carol: This last year we gathered leeks, but they are so strong. What do you do with them? I make wild onion soup, do you do leeks the same way?

Irv: Yes, the same thing. But after they've been cooked, they are just like greens, like spinach. They're not strong. It's the raw ones that are strong.

Carol: Out at Joses's the leeks are like a green carpet in the woods.

Irv: I drive thirty miles to get leeks. Some people think it's too much work. It's time consuming, you have to gather it, wash it, and cook it. It's much easier to go the store and buy it in a can. But when you're harvesting, you're into the dirt and you can feel that. It's a much better feeling.

Carol: I think the Creator is there.

Irv: Yes, yes, and the first time you go out, you give thanks for each one as you gather. The leeks, the wild onions. We have wild onions from last spring, we cut them up and put them in soup. We had wild onions in our mashed potatoes last night.

Carol: Do you freeze them?

Irv: Yes, freeze them and then when you get ready you just take a package out and cut them up and throw them in the soup.

Carol: My experience picking wild onions on Cornell property, hearing sirens, but no one bothered me.

Irv: There was an estate sale up in . . . Helen and I went to and as we crested this hill it was just green with leeks. Right alongside the road and we said we're going to come back. So we got a trash bag and came back to pick leeks, knowing we were on private property. Cars going by would slow down. It was right on a crossroads. I knew sooner or later somebody's going to come down. Shortly a lady came down and said, "Are you aware you're on private property?" I said, "Yes, I knew someone would come, so I could ask permission. I didn't know who owned the property. We're picking leeks." She said, "Well, you know I've got wildflowers here and I don't want them disturbed." I said, "I haven't destroyed anything. If you look closely, you see what I have here, probably, I have almost three-quarters of a trash bag full of leeks which I have dug up. Look around, now, can you tell where I dug?" She looked around and said, "I can't tell." Then, I said, "Your wildflowers are still there." I said, "That's the difference between my people and your people. I come through this area and you can't even tell where I've dug." So I said, "I have not removed anything else, and I've hardly disturbed the land. All I took was leeks." Well, she said, "I

owe you an apology. As long as I own the land you can pick here."
So every year I go there and every year she's looking for me. I
always go up to her house and knock on the door. She says, "We've
been waiting for you. We got lots of leeks."

Carol: Now, you're part of the natural cycle.

Irv: Yes, tradition.

Carol: I never thought about that, but when I pick I only take
from where the leeks or onions are thick.

Irv: Yes, it grows in clumps, so I take a shovel and get the big
clumps. But where there's singles, I leave those to seed because
next year they will be clumps.

Carol: That makes sense.

Irv: It's like the buffalo. The buffalo will not clear out an area
in their natural environment. They'll always leave some growing.
Whereas, domesticated animals, cows, eat everything.

Carol: Yes, just like the heron. I always wondered why the
fishermen don't pay attention to when the heron fishes and that's
when the fish are there. When did things change from corn soup as
daily meal to special occasions?

Irv: With the invention of the car. Somewhere around the late
1940s it changed. Up until that time, our people stayed mostly on
the reservation. They didn't go into town very much. My grandfather went into town to . . . store-bought clothes. We were talking
about this the other day and . . . bakers bread. I remember TipTop
bread, all sliced white bread. So anybody who had bakers bread
were the elite because they had purchased bread rather than bis-
cuits and hosgwe. . . . When electricity and when cars and trucks
came into the territory, it changed.

Carol: The lifestyle changed with the modern world. It takes
a lot of time to live the old way.

Irv: Yes, but you didn't work out then. You were farming and
so forth. And you took your farm stuff down to sell to buy shoes or
whatever. But that was the only transaction. Most of the food was
home canned goods. All the homes had fruit cellars. All the homes
had cellars with rows and rows of home canned foods because you
were canning all of your food. They had chickens and pigs. They
butchered cows and pigs. They were hunting deer, pheasants, rabbits. They were trapping. They had muskrats . . . raccoons . . . they

sold their furs. And besides that they had crafts. Because people have always been looking for drumsticks, canes, chairs. Bows and arrows have always been a fascination for non-Native people who would come out and buy a bow and arrow. . . . And this was how they made a living.

It wasn't until after the Depression that people started going to the city to get jobs and started getting cars. And once that happened they went away from the farming and started going into the city. And, in the city, they started using store bought goods and food. Now, you've got families who buy bags and bags of canned goods and they don't eat fresh vegetables.

Carol: So, it's continued by some families and not by others?

Irv: That's always been present. There's always been a core of people who've stuck to the traditions. That's why the language is still here and that's why the ceremonies are still here. There might have only been a few, but it was still going on. Those few were able to pass it on to their children so that it continued. It's still here, and we look around, and it's still here.

Appendix 6

Interview of Katsi Cook, Akwesasne Mohawk Midwife, by Carol Cornelius, June 6, 1992

Carol: I'd like to talk with you about an article for young people called "Corn in My Kitchen."

Katsi: The best way to present corn to the young people is to have them look at some of the material culture related to the corn. I think, at least for me, I never really understood the corn until I started planting it and growing it and looking for the metaphors and analogies of it in the culture. Even in the corn washing basket making, how the weave is, a basket maker would probably be better at talking about it than I am. The one back home who makes it, the weave is of course to help get the hull off the kernel, but the corn washing basket itself is to me an interesting aspect of corn culture.

I guess as a midwife, I really never had the depth of understanding before about traditional roles of Iroquois women until I grew corn. And I never understood some of the depth of understanding of knowledge of science in traditional societies. That the moment you begin to work with corn, even just in harvesting, everybody picks an ear and pulls back the husk. It's like seeing a brand new baby the way everybody reacts. "Look at this one, here's one red kernel among all white."

It's an interesting education in genetics. I know John Mohawk used to say that the power of the Iroquois women came through their control of agriculture. That's the fields came under the domain of the women. When you grow a field of corn you get a sense of that. He said that the women controlled the sexual economy of the village because they decided who would mate, who the couples would be that produced children. Even though that may sound a little cold and calculating in a society that's overdosed with notions of romantic love. It was very interesting because when you read the Jesuit journals and early explorers into the continent, they all talk about how healthy Native People were, how strong they were, how resistant to infection. When I went to the Smithsonian to research traditional childbearing I read stories from all over the continent and army surgeons who had gone out into the Indian communities were amazed at the resistance to infection that Native women had. There are stories in Northern Canada of a husband and his wife being out in the bush and there being problems with the delivery and a cesarean was performed and the woman survived. I believe this was due to the resistance to infection. When you look at what it would involve I can see where there would be survival rate from cesarean hundreds of years ago before it ever became part of Western medicine.

So the strength of women is an overwhelming theme you read in any history of Native women and I believe it goes back to the corn. Even in looking at midwifery one of the things we hear in oral stories is that if there was any problem with the baby the Grandmother would take that child home with her. It was up to her to decide that child's future. If the child wasn't able to survive physically in that society the child would be put to death, or allowed to die. An infant will die in a relatively short time if it's not nurtured. It's almost an animalistic thing, and birth is an animalistic thing. I think the women got their sense of how to do this, or their strength to do it because they were in control of the genetics of the community, if in fact they were the ones who controlled the sexual economy. They wouldn't allow marriage for individuals who had obvious deformities, and in the same way the selection of the seed was in choosing characteristics they wanted to reproduce.

I got a lot more sense of our culture in working with corn. The notion of "corn in my kitchen" is a really wonderful one because it brings home to us something that is very complex and in every part of our lives, our spiritual lives, even our emotional lives. One of my favorite teachings about the corn is, that corn is where we learn midwifery. You see corn at birth, at puberty, as a symbol of fertility, reproduction, abundance. So corn has a great deal of meaning. On an emotional and spiritual level corn also has great meaning. We know this because the short-eared corn has kernels covering the nose. The middle one of that is used for the survival of the infant if the mother has died in childbirth. It's ground up in water and fed to the baby to remind the baby even though its lost its mother, it still has its mother the corn. That babies when they are born, the powder that was used, it wasn't Johnson and Johnson, it was ground up white corn and the baby was cleansed with that after birth. I think that's what they did instead of bathing the child in water although the two are not contradictory.

So corn goes into every facet of our lives once you really begin to look at it. I guess what I realized in getting involved in the cycle of the corn, the blessing of the seed, the preparation of the field, the ceremonies during the corns gestation, and the harvest, and storage, and care of the seed is that this is reproduction, this is pregnancy this is the metaphor, all the ceremonies all the songs, everything we do is what we're supposed to do at birth. Raising corn is where we learn midwifery. So the same knowledge is applied. The corn's gestation, for example, and Jorge Quintana gives a really good talk about this, the similarities between the gestation of the corn and human gestation. It's a shortened time period for the corn, but some of the same growth factors are there, instead of a nine month gestation, it's a three month growth period. Look at the corn's reproduction, the fallopian tube of the woman or the seminal vesicle is analogous to the corn silk which directs pollen to where the kernel grows. The corn silk is used in pregnancy to help prevent or deal with bladder infections, which happen to be one of those things that plague pregnant women because hormones influence their tubes from the kidney to the bladder. Every part of the corn was used as medicine.

A traditional Mayan priest was the one who gave me a deeper understanding of the relationship of the gestational cycle of the corn to that of women and of a community. He said when you go to plant a field of corn you ask the permission of the insects and all the life in that particular field because you're going to be destroying some of that to make way for corn to grow. In the same way, a couple that wants to have a baby should make that same preparation or plea to the spirit world to assist them in their endeavor to bring a new life from the spirit world to this world and to help them in the growth of that.

The green corn dance, those ceremonies that are done during the corn's growth cycle are also analogous to ceremonies that need to be done during a woman's pregnancy. There are ceremonies that should be done in each trimester, each growth cycle of the fetus. It may differ with each woman but the themes are the same. To connect her with a nurturing elder, and the corn certainly teaches us nurturing and that's where it becomes even an emotional teaching for us. So the permission is asked to create the field. The relationship between the corn, beans, and squash, the Three Sisters, the companionship, again, a nurturing relationship is what helps the corn to grow. Then the ceremonies during its growth and then at harvest time.

The story of the corn is one of my favorite stories that I think corn in my kitchen should include and that Roderico told us. He said, when a new baby is born the people in the home are supposed to be very happy and people are supposed to talk to each other like you want the baby to talk. Teach the new baby by your actions. In a home where there's a new baby you don't want any loud noises or angry people, or bad vibes, the same, he said, is with the corn because people used to hang it in their homes. They didn't have barns like we do now. They would store the corn in caches in different places in nature, but it made sense to keep some of it hanging from the rafters in the kitchen. People in Mexico still do that. They have a way of saying, well, three ears of corn a day for each person in the family and they'll hang that many ears from their rafters. They said when that corn is hanging in your kitchen its the same as when there's a baby in the house. You have to talk nice to each other. You have to be respectful of one another. You can't use any angry language or bad words because the corn can hear you. If it hears all this negativity it won't grow for you the next year. I think that kind of metaphor between the corn and the children is very important, certainly in the absence of corn and babies everyone should strive to have peace in their homes. That's what we all need, to be nurtured. Again that theme of nurturing in pregnancy, in the family, and in the growth of the corn is something that corn teaches us. It teaches us how to take care of each other. No wonder they say return to the arms of mother corn, corn is our mother in the same way that the earth is our mother, but it's more immediately so our mother because that's what feeds us. We actually take it into our bodies so corn is a very important thing. I believe in raising a traditional family, a family that has values of closeness, of nurturing each other because through the physical activity of doing something you learn. You don't learn just by being told to do this. You learn by doing it. Sometimes only the physical experience of something—they say See One, Do One, Teach One, and that's the basis of empirical training. I think corn really does teach us in many ways. Again, the spiritual, emotional, and physical.

So corn in my kitchen I'd like to see focus on preparing the corn because it is an involved project, but it gets you out in nature. It makes you have to look for the hardwood ashes which release those amino acids from the hull so it can be used by the body. Corn is a great teacher in just the chemistry of food. You have to know how to use the corn. Luckily for me there are simple ways to use the corn without having to go to all the bother of washing it. It's very time consuming and I don't always get to wash corn. My favorite is corn mush, roasting the corn in the pan and sitting there sorting the corn and the conversations that go on when you are able to do that. To take time out to create the foods that corn provides is a time consuming thing, but there's much more value in it than opening up a box of pizza that's for sure.

Carol: I can remember too, when you used the metal hand grinder that attaches to the table, one would be sorting, one parching, and one grinding, and we would take turns.

Katsi: It was a cooperative effort, that's for sure. I use a blender instead of the hand grinder when I'm alone. It's almost like a meditation because you have to keep all the parts straight in your mind, you can't take short cuts.

Carol: It's like that corn basket. You started with the corn basket talking about the weave and all those different components, I'm looking at the basket in the corner there, there's your spiritual, your emotional, the actual physical activity, planting, all woven into an organic whole.

I was telling a woman about this wonderful electric corn grinder you have in the shed, but looking at that, on the one hand it's a modern technology that saves us time and energy but we sacrifice the interaction of the family working together.

Katsi: We never use that electric grinder when we're making corn for the household. That's for IPNC. We bought that thinking we're going to mass produce this stuff, but in the absence of that we don't even use it. You've got to have a lot of corn moving through it. If we were going to make two big trash cans full of roasted corn I would use that. It's so easy to whip up a batch of roast corn mush. I use a blender, I even use my coffee grinder now. It's easy. If you make something too long people don't say they want it because their expectation is immediately they have to get the grinder out, put it on the table and so on. And with families today we really don't have the time. But there are those special times when it's a cold day out in the fall and everyone hungers for the taste of that roast corn. It's never too much work to sort it out, roast it on the stove and then grind and cook it. It's not that hard. When I think of all the different ways of fixing corn, that's probably the most attractive for kids and easiest for them to implement. Once you have mush you can make a lot of things with it. I tried making about a dozen jars of mush so I'd have it handy, but I always end up giving it away because it's a gift people love having.

Carol: I find that with canning too.

Katsi: Sure, because you're giving them your time and your love and all the thoughts you had when you were making it. It's not like a can you pull off the shelf at Wegmans. The older I get the more I appreciate that. I remember my grandmother canning, freezing, raising her own meat. I used to wonder, why does this old woman do this. All she's got to do is go to the store and buy these things. Then when I became a grownup and figured out how much things cost and you're not just paying for the meat you're paying for guy that raised it and the guy he sold it to market it and all the chemicals he put in the feed to keep it healthy and how those chemicals react in your body. It's a good argument for growing your own, for producing as much as you can of your own. I think in this age we have to be more aware of our consumer behavior because we need to protect the earth. These environmental problems we read about and hear about on television are not anyone's problem but our own. It's easy to think, well the town landfill is being shut down that's not my problem, well it is. The simplest behavior like throwing a piece of paper out of your car makes a difference to society. It's growing incrementally. They're saying right now for every chemical factory that exists in ten more years there will have to be ninety, if it keeps growing in the increments it is right now. You hear people in the chemical industry saying we need new and improved chemicals. Imagine that every place there's a chemical factory now there being ninety more. I don't think the earth can sustain it. But with the population growing, it's even as private as deciding how many children you have and how you're going to raise those children with what expectations. So the kids growing up today really need a way to think seriously about these things in their own lives.

Carol: Isn't it interesting that they see the solution as more chemicals?
Instead of asking how things worked before.

Katsi: Yes. It's only individuals who start educating themselves about the issues. One of the arguments for saying we need

more chemicals is there are these new diseases, but one pharma-
cologist I know says everything we need nature has already, it
already exists. The development of new chemicals is manipulating
genetic materials in the lab and that's what frightens me.

That's the lesson of corn, think of how much went into taking
corn from a grass. To just try to comprehend what level of human
organization and intellect it took to take that corn from grass to
corn that is today the economic base of the planet. Without corn
we'd have some serious crisis right now. Even though there's a lot
of starvation on this planet, it would be even worse without corn.
I'm amazed when I think of what corn really means to society. This
one gift of Native People to the world, how it's been developed and
used. Again, corn itself is metaphor for how we look at nature and
how we use science to the betterment of human beings and the
planet, not just dollars. That's something the children really need
to think about into the future.

Carol: You have the whole project going with corn, the big
field out there, people coming all the time especially for the harvest, your own gardens and experimental gardens in the gift of the
Grandfather corn. How does all of that affect your family, you're
doing a family in the midst of all of that, how does that affect these
gorgeous kids?

Katsi: I think that corn is the best teacher. It's almost like
having a college out here that we're part of. I get a lot of joy from
the gardens and the work that's being done here, because, number
one, we're not doing it all ourselves, we're not doing it alone. We
get to share it with other families who have the same and similar
interests. There is a cultural base to it. It's not just an agricultural
project that Cornell's doing because they want to figure out how to
get better yields or how not have to use this kind of chemical. It
really does affect us on all of those levels I talked about before, the
physical, the emotional, the spiritual. It's been a real teaching for
our family because it teaches us about sharing, sharing time, sharing space, sharing your spiritual understanding of what life is
supposed to be.

I use to hang the corn in my kitchen, then we got an infesta-
tion of that moth, that Iroquois moth, maize moth. I thought I
better not hang it in my kitchen until I get rid of this moth. But,
I love it, the experience of being there for the seed blessing, the
planting, and watching the corn grow. It really reminds us and
gives us a sense of the cycle of life. A constant reminder. It's almost
like you gauge time by how high the corn is. You pay better atten-

tion to the weather and the lunar cycles. You just have to pay better attention to the environment when you have something growing. With corn, because it has special meaning to us it makes it that much more strong. The harvest is such a joy because we have visitors we know, or people will bring friends that we didn't know before and we'll get to know new people. So we have a very active life out here even though we're not in the city. It can get pretty boring out here because there's no mall, no stores, but there's never a lack of things to do. I experience it that way and I think it's been a very positive thing for our family and it really does bring the family closer together.

Carol: Even the air is different out here. At night you can see the stars. I feel like you've woven a basket for me. Each splint, each strand of that, when you weave it together you wind up with a whole, that's strong. That's the sense I'm getting. When I talk to children in schools I show my corn basket and explain to them that I've tried metal and plastic colanders, sieves, there's nothing that can replace this corn basket that modern man has invented. Nothing can replace the corn basket. It's meant for a purpose and it does that the best.

Katsi: Well, the basketmakers back at Akwesasne had almost forgotten how to make one because so many people stopped growing corn. And now, there's really only two serious corn growing projects up there. But, it's catching on more and more because people love to eat it. That hasn't disappeared. The ability to get it has changed, the overwhelming growth was gone once they put the Seaway in.

Carol: I always remember the Richmond Sisters, one time I gave them a ride home. They pointed out that field across from Tom Porters's and said they could remember when that was a huge field of corn, and at that time it was weeds. Today, its a bingo hall or casino. I keep thinking about the history of that one field. Who's raising corn now?

Katsi: There's Wally Ransom, and Mark Narcissan's family on Cook Road. Marie Cook's farm. There's always smaller families raising corn. Wally showed me his two huge fields. I remember down on Sugarbush Island, Joe and Hattie Mitchell they used to grow corn like crazy and they were elderly people. Once you have an attachment to it, it's almost like life doesn't have any meaning without it. You have a sense of being lost without it.

Carol: Like me, this year, I'm not planting because I'm moving and there's a piece of me missing. It's like you haven't gone out there and done this thing which you're supposed to do with the Creator.

Katsi: That's right. That's a good way of describing it. Almost like it's a part of your mental, emotional, and physical state of health that you have do this. It's almost instinctual. If I don't do this I'm not going to feel very good.

Carol: The pattern I'm finding is this growing of corn in families from generation to generation, like you mentioned Joe and Hattie, they grew up with corn and they continued raising it. Other people in the community depended on them to raise the corn that's why the eating and enjoyment of it hasn't diminished because people can still get corn. What we've seen is some have passed on or they're too old now and in some communities there's no one raising corn. Or there's people trying to raise five or six rows for their family but they depend on the Rickarts or Wally or this project to be like the corn granary.

Katsi: That's right. They depend on this project to get the seed. I've taken seed out to older people and they are so happy to get it. They grow it and use what they grow.

Carol: I've heard people complaining about raccoons when they plant only five or six rows.

Katsi: You don't worry about raccoons when you've got three or four acres, they can have their share.

Carol: That's what I thought, if you plant five or six rows, you've planted enough for the raccoons. If you plant twenty rows you have five rows for them and fifteen rows for your family. I've seen mild mannered people who would turn into vicious killers if they could get hold of that raccoon! I think it's a loss of that understanding that we have to plant enough so the animals have their share. I learned that from picking corn here. If you pick in the first five rows you don't get much, but if you pick in the middle you get a lot of corn, there's good corn in there. That thrill of discovery when you husk an ear. Do you know of anyone who still plants in hills?

Katsi: Because of machinery, if there's a lot of corn, an acre or more, nobody plants in hills. I think the women should keep it up to express that relationship with the corn, beans, and squash, the Three Sisters. Most of us have sisters, if not actual biological sisters, we have women whom we relate to. I think that's a very woman-oriented part of our culture and there should be Three

Sisters gardens planted by the women just for that purpose. The squash could have more of a ceremonial purpose, the gourd squash used for rattles, those are part of the Iroquois culture that needs to be revived. I imagine they would use those gourd rattles when they would sing to the women in her menses, in the moon lodge, separate from the village. Different aunties or grandmothers would go sing songs to her and talk with her so that by the time she was ready to become a wife and a mother she already knew about her sexuality and taking care of babies. Not only from her experience as a young woman taking care of children, but the women would instruct her explaining what you do. There were a lot of instructions necessary because back then abstinence was a traditional part of sexuality. When a woman had a baby, or was menustrating, she wasn't around men. You have a real strong sense of women's culture, not like the feminists talk about, it wasn't anti-man, it was very nurturing of the individual woman because by being nurtured you learn how to nurture. That's the biggest lesson Native Women have to share with other women. That we need to nurture one another.

Carol: I liked the way Roderico explained that in his article, that just as the earth rests after harvest, a woman has to rest after giving birth. I can't help but notice you have some framed Ernest Smith paintings. You were talking about the gourd rattles, Ernest Smith has those in his paintings.

Katsi: It's a calabash gourd. I was amazed to read your proposal and find out there were 240 paintings. I've only ever seen six of them.

Chapter Notes

Chapter 1   The Problem: Stereotypes

1. Henry, Jeannette, and Rupert Costo. *Textbooks and the American Indian.* California: American Indian Historical Society, 1970: 2–3.

2. Allport, Gordon W. *The Nature of Prejudice.* Garden City, N. Y: Addison-Wesley Publishing, 1958: 187.

3. Allport: 196.

4. Council on Interracial Books for Children. *Stereotypes, Distortions, and Omissions in U.S. History Textbooks.* New York, 1977: 131–133.

5. Hirschfelder, Arlene B. *American Indian Stereotypes in the World of Children.* Metuchen, New Jersey: Scarecrow Press, 1982: 51.

6. Stensland, Anna Lee. *Literature by and About the American Indians.* Urbana, Illinois: National Council of Teachers of English 1973: 9.

7. Harris, Helen L. "On the Failure of Indian Education." *The Clearing House* 48 (December 1973): 242.

8. Harris: 245.

9. Hoxie, Fedrick E. *"The Indian Versus the Textbooks: Is There Anyway Out?"* Newberry Library, 1984: 21.

10. Stensland: 10.

11. Harris: 242.

12. Hoxie: 20.

13. Vogel, Virgil J. "The American Indian in American History Textbooks." *Integrated Education.* Vol. 3, May–June 1968: 22.

14. Hirschfelder,1982: 20–24.

15. Stensland: 8.

16. Hirschfelder, 1982: 12.

17. Hirschfelder, 1982: 23.

18. Berkhofer, Robert F. *The White Man's Indian: Images of the American Indian from Columbus to the Present.* New York: Random House, 1978: 71.

19. Rouse, Linda P., and Jeffery R. Hanson. "American Indian Stereotyping, Resource Competition, and Status-Based Prejudice." *American Indian Culture and Research Journal*: 15, 3, (1991): 5.

20. Rouse and Hanson: 15.

21. Berkhofer: 91.

22. Elson, Ruth. *Guardian of Traditions: American Schoolbooks of the Nineteenth Century*. Lincoln: University of Nebraska Press, 1964: 73–74.

23. Henry: 7–9.

24. Council on Interracial Books for Children, 1977: 67.

25. Ferguson, Maxel J., and Dan B. Fleming. "Native Americans in Elementary School Social Studies Textbooks." *Journal of American Indian Education*: 23 (2) January 1984: 10.

26. Hoxie: 1–2.

27. Hoxie: 24–26.

28. Hoxie: 28–30.

29. Charles, James P. "The Need for Textbook Reform: An American Indian Example." *Journal of American Indian Education*: May1989: 3–7.

30. Hirschfelder, 1975: 34.

31. Hirschfelder, 1975: 34–35.

32. Hirschfelder, 1975: 37.

33. Hirschfelder, 1975: 38.

34. Hirschfelder, 1975: 38.

Chapter 2   Cultural Evolution:
The Theories behind the Stereotypes

1. Pearce, Roy Harvey. *Savagism and Civilization*. Baltimore: Johns Hopkins Press, 1953: 24–25.

2. Pearce: 20–21.

3. Pearce: 19. See Starna in Campisi.

4. Pearce: 81–82.

5. Berkhofer, Robert F. *The White Man's Indian: Images of the American Indian from Columbus to the Present*. New York: Random House, 1978: 36.

6. See: Berkhofer, Pearce, Kuper.

7. Pearce: 66–67.

8. Cook, Frederick. *Journals of the Military Expedition of Major General John Sullivan Against the Six Nations of Indians in 1779, with Records of Centennial Celebrations*. Auburn: Knapp, Peck & Thomson, 1887.

9. Usner, Daniel, H. "'We do not view it so': Iroquois Livelihood in the Face of Jeffersonian Agrarianism."*The John Ben Snow Lecture Series on Iroquois Life and Culture*, ed. Barbara J. Blaszak, Syracuse, NY: Le Moyne College, 1987: 31.

10. Usner: 25–26.

11. Berkhofer: 161.
12. Sale, Kirkpatrick. *Conquest of Paradise.* New York: Alfred A. Knopf, 1990: 74–75.
13. Sale, Kirkpatrick: 78.
14. Sale, Kirkpatrick: 40.
15. Sale, Kirkpatrick: 81–82.
16. Sale, Kirkpatrick: 79.
17. Kuper, Adam. *The Invention of Primitive Society* New York: Routledge, 1988: 8.
18. Kuper, Adam: 7.
19. Kuper, Adam: 8.
20. Kuper, Adam: 8.

Chapter 3   The Theories Become the Standard Curriculum

1. Goodenough, Ward H. "Culture, Language, and Society." *Addison-Wesley Module in Anthropology,* 1971: 17.
2. Goodenough: 18.
3. Stocking, George W. "Franz Boas and the Culture Concept in Historical Perspective." *American Anthropologist,* 68 (1966): 870.
4. Stocking: 871.
5. Rouse, Linda P., and Jeffery R. Hanson. "American Indian Stereotyping, Resource Competition, and Status-Based Prejudice." *American Indian Culture and Research Journal:* 15, 3, (1991): 16, fn.
6. Berkhofer, Robert F. *The White Man's Indian: Images of the American Indian from Columbus to the Present.* New York: Random House, 1978: 62.
7. Berkhofer: 63.
8. Berkhofer: 64.
9. Berkhofer: 64.
10. Berkhofer: 66.
11. Berkhofer: 183.
12. Posner, George, *Analyzing the Curriculum.* New York: McGraw Hill, Inc., 1992: 10.
13. Posner, George, and Alan N. Rudnitsky. *Course Design, A Guide to Curriculum Development for Teachers.* White Plains, New York: Longman, Inc. 1986: 206.
14. Posner: 11–12.
15. Vallance, Elizabeth. "Hiding the Hidden Curriculum: An Interpretation of the Language of Justification in Nineteenth-Century Educational Reform." Giroux, Henry, and David Purpel, eds. *The Hidden Curriculum and Moral Education.* Berkeley, California: McCutchan Publishing, 1983: 14–15.
16. Vallance: 16–20.

17. Kliebard, Herbert. *The Struggle for the American Curriculum 1893–1958*. New York: Routledge & Kegan Paul, 1987: 126.

18. Kliebard: 126–128.

19. FitzGerald, Frances. *America Revised—History of Schoolbooks in the Twentieth Century*. Little Brown & Co., 1979.

20. Banks, James A., and James Lynch. *Multicultural Education in Western Societies*, Westport, Connecticut: Praeger, 1986: 62.

21. Hoxie, Fedrick E. "The Indian Versus the Textbooks: Is There Anyway Out?" Newberry Library, 1984. *Social Studies* (tentative Syllabi) K–12. The University of the State of New York, the State Education Department, Bureau of Curriculum Development, Albany, New York, 1987.

22. Banks: 192.

Chapter 4   Valuing Diversity through Culture-Based Curriculum

1. Barreiro, Jose. "First Words." *Northeast Indian Quarterly*. Summer 1991: 3.

2. See: Taylor, Charles. "Interpretation and the Sciences of Man." Eric Bredo and Walter Feinberg, *Knowledge and Values in Social and Educational Research*. Philadelphia: Temple University Press: 153–186. Geertz, Clifford. *The Interpretation of Cultures*. New York: Basic Books, 1973. Sosa, John. The *Maya Sky, The Maya World: A Symbolic Analysis of Yucatec Maya Cosmology*. Diss. SUNY-Albany, 1985. Ortiz, Alfonso. *The Tewa World*. Chicago: University of Chicago Press, 1969.

3. Wilson, Gilbert L. *Buffalo Bird Woman's Garden*. St. Paul: Minnesota Historical Society Press, 1917. Rpt. 1987: xviii.

4. Sosa: 116.

5. Geertz, 1973: 127.

6. Ortiz, Alfonso. In Beck, Peggy V., and Anna L. Walters. *The Sacred Ways of Knowledge, Source of Life*. Tsaile, Arizona: Navajo Community College, 1977: 5.

7. Ortiz. In Beck and Walters: 5.

8. Vecsey, Christopher. *Imagine Ourselves Richly, Mythic Narratives of North American Indians*. New York: Crossroad Publishing, 1988: 23–27.

9. Ortiz, Alfonso. *New Perspectives on the Pueblos*. Albuquerque: University of New Mexico Press, 1972: 137.

10. Vecsey: 29.

11. Weatherford, Jack. *Indian Givers—How the Indians of The Americas Transformed the World*. New York: Crown Publishers, 1988.

12. Sosa: 479.

13. Gonyea, Ray. Introduction. By Fred R. Wolcott. *Onondaga: Portrait of a Native People*. Syracuse: Syracuse University Press, 1968: 11–32.

14. Hammell, George "The Prehistory of the Seneca Iroquois in the Genesee Valley." *A Genesee Harvest*: 31.

15. Ortiz: 5.

16. *Sun Tracks, Between Sacred Mountains, Navajo Stories and Lessons from the Land.* Tucson: Sun Tracks and the University of Arizona Press, 1982.

17. Skinner, Linda. "Teaching Through Traditions: Incorporating Native Languages and Cultures into Curricula." *Indian Nations at Risk: Listening to the People.* Washington, D.C.: U.S. Department of Education, 1992: 56.

18. Posner, George. *Analyzing the Curriculum.* New York: McGraw Hill, Inc., 1992: 38.

19. Joyce, Bruce, and Marsha Weil. *Models of Teaching.* Englewood Cliffs, New Jersey: Prentice-Hall, 1986: 42–43.

20. Lincoln, Yvonna S., and Egon G. Guba. *Naturalistic Inquiry.* Newbury Park, California: Sage Publications, 1985: 205.

21. Ogawa, Rodney T., and Betty Malen. "Towards Rigor in Reviews of Multivocal Literatures: Applying the Exploratory Case Study Method." *Review of Educational Research*, Fall 1991, Vol. 6:265.

Chapter 5   The Thanksgiving Address: An Expression of Haudenosaunee World View

1. Ortiz, Alfonso. *New Perspectives on the Pueblos.* Albuquerque: University of New Mexico Press, 1972.

2. Sosa, John. *The Maya Sky, The Maya World: A Symbolic Analysis of Yucatec Maya Cosmology.* Diss. SUNY-Albany, 1985: 116.

3. Chafe, Wallace. *Seneca Thanksgiving Rituals.* Bureau of American Ethnology, Bulletin 183, Smithsonian Instituion, Washington, DC: U.S. Government Printing Office, 1961: 6.

4. Foster, Micheal, F. *From the Earth to Beyond the Sky: An Ethnographic Approach to Four Longhouse Iroquois Speech Events.* Ottawa: National Museums of Canada, 1974: vi.

5. Chafe: 1.

6. Hewitt, J. N. B. *Iroquoian Cosmology.* Annual Reports Bureau of Ethnology, Part 1, 21st report 1899–1900, Part 2, 43rd report, 1925–1926: 570.

7. Hewitt: 606.

8. Chafe: 7.

9. Morgan, Lewis Henry. *League of the Ho-De-No-Sau-Nee or Iroquois,* New York: Burt Franklin, 1850. Rpt.1901: 210–213.

10. Chafe: 29–31.

11. Foster: 324–331.

12. Chafe: 17.

13. Foster: 292–293.

14. Foster: 294.

15. Chafe: 23.

16. Foster: 299–300.
17. Chafe: 21.
18. Chafe: 29.
19. Foster: 322–323.
20. Foster: 329–330.
21. Chafe: 29.
22. Foster: 333.
23. Chafe: 33.
24. Foster: 338.
25. Chafe: 35.
26. Foster: 339.
27. Chafe: 37–39.
28. Foster: 345.
29. Chafe: 41.
30. Foster: 349.
31. Chafe: 33.
32. Chafe: 41.
33. Foster: 351.
34. Foster: 356–358.
35. Chafe: 43.
36. Hewitt: 161.
37. Hewitt: 163.
38. Hewitt: 166.
39. The Fire Dragon is not in the Onondaga version, but is in the Gibson version recorded in 1900 and in the Seneca version. Hewitt: 224, and Cornplanter: 13.
40. Cornplanter, Jesse J. *Legends of the Longhouse.* Philadelphia, Pennsylvania: J. B. Lippincott, 1938: 17.
41. Hewitt: 453–608.
42. Parker, Arthur. *Parker on the Iroquois.* Edited with an introduction by William N. Fenton, Syracuse University Press, 1968. Wallace, Paul, A.W. *The White Roots of Peace.* Philadelphia: University of Pennsylvania Press, 1946. Rpt. Saranac Lake, New York: Chauncy Press, 1986.
43. Parker: 105, Wallace: 39.
44. Vecsey, Christopher. *Imagine Ourselves Richly, Mythic Narratives of North American Indians.* New York: Crossroad Publishing, 1988: 114–115.
45. Parker, 1968, Book II: 10–11.
46. Parker: 40–42.
47. Parker: 51.
48. Parker: 54.
49. Parker: 54.
50. Parker: 38.
51. Parker: 67–68.
52. Parker: 74.
53. Parker: 64.
54. Parker: 66.

55. Parker: 137.
56. Parker: 68.
57. Parker: 7.
58. Parker: 7–8.

Chapter 6   Corn as a Cultural Center of the Haudenosaunee Way of Life

1. Hewitt, J. N. B. *Iroquoian Cosmology*. Annual Reports Bureau of Ethnology, Part 1, 21st report 1899–1900, Part 2, 43rd report, 1925–1926.

2. See: Morgan, Lewis Henry. *League of the Ho-De-No-Sau-Nee or Iroquois*. New York: Burt Franklin, 1850. Rpt 1901. Chafe, Wallace. *Seneca Thanksgiving Rituals*. Bureau of American Ethnology, Bulletin 183, Smithsonian Instituion, Washington, DC: U.S. Government Printing Office, 1961. Foster, Micheal, F. *From the Earth to Beyond the Sky: An Ethnographic Approach to Four Longhouse Iroquois Speech Events*. Ottawa: National Museums of Canada, 1974. Cornelius, Carol. "The Thanksgiving Address: An Expression of Haudenosaunee Worldview" Ithaca, New York. Akwe: Kon Journal, Fall 1992: 14–25.

3. Morgan: 175. The Thanksgiving Dance, or Drum Dance, which Wallace Chafe attended was held at the Tonawanda longhouse on February 6, 1960, as part of the MidWinter ceremonies. He provides a detailed description of the dance which alternates the spoken thanksgiving with dances. In the Drum Dance, additional sections are included which express gratitude or thankfulness for the chiefs, faithkeepers (those men and women delegated to insure the continuance of the ceremonies), and the children. The Thanksgiving Address is one of the Four Rituals considered the most important in the ceremonies: 1) the Great Feather Dance which honors the Creator, 2) Thanksgiving Dance or Drum Dance, 3) the Peach Stone Game, which originated in the cosmology and 4) personal chants.

4. See: Morgan: 175–216. Hewitt. Parker, Arthur C. *Iroquois Uses of Maize and Other Food Plants*, Albany, University of State of New York: New York State Museum Bulletin, No. 144, 1910. Smith, Erminnie A. *Myths of the Iroquois*. U.S. Bureau of Ethnology, 2nd Annual Report, 1880–81. Washington D.C. 1883.

5. See: Parker, Arthur C. "Secret Medicine Societies of the Seneca." *American Anthropologist*, New Ser. 1909, v.2, No.2: 161–185. p. 179. Parker, 1910: 272, 179.

6. Parker, 1910: Fig.2: 28.

7. Harrington, M. R. "Some Seneca Corn-Foods and Their Preparation." *American Anthropologist*, N.S. 10, 1908: 575–590.

8. Parker, 1910: 37.

9. Parker, 1910: 36. Morgan: 191.

10. Cornplanter, Jesse J. *Legends of the Longhouse*. Philadelphia, Pennsylvania: J.B. Lippincott, 1938: 18–19. Parker, Arthur C. *Seneca Myths*

and Folktales. Buffalo: Historical Society 1923, Rpt. University of Nebraska Press, 1989: 64.

11. Parker, 1923: 64.

12. Morgan: 153. Beauchamp, William, M. *Civil, Religious and Mourning Councils and Ceremonies of Adoption of the New York Indians*. New York State Museum Bulletin 113, 1907: 195.

13. Cornplanter: 16.

14. Parker, Arthur C. *The Constitution of the Five Nations*. New York State Museum Bulletin No. 184, 1916: 76–82.

15. Curtin, Jeremiah, and J. N. B. Hewitt. *Seneca Fictions, Legends, and Myths*. 32nd Annual Report of the Bureau of American Ethnology, Smithsonian Institution, 1910–11: 273–276. Parker, 1923: 386–393.

16. Curtin: 274.

17. Parker, 1923: 389–392.

18. Beauchamp: 196.

19. Curtin and Hewitt: 647.

20. Curtin and Hewitt: 647–649.

21. Parker, 1923: 205–207.

22. Jenness, Diamond. *The Corn Goddess*. Ottawa, Canada: National Parks Branch, National Museum of Canada, Bulletin no. 141, Anthropological Series no. 39, 1956. Curtin and Hewitt: 636–642.

23. Curtin and Hewitt: 701–704.

24. Curtin and Hewitt: 649–653.

25. Parker, 1910: 39.

26. Parker, 1910: 37.

27. Smith, Erminnie: 59–61.

28. Parker, 1923: 108.

29. Curtin and Hewitt: 474–481.

30. O'Callaghan, E. B. *The Documentary History of the State of New York*. Vol. I, Albany, New York: Weed, Parson & Co., 1849: 11–13.

31. Morgan: 307–308.

32. Morgan: 23–24. Morgan indicates that in the 1840–1850s, there were also single-family houses spread over a large area. Morgan cited population figures available to him: LaHantan in 1735 (70,000), Coursey in 1677 (15,000), Bancroft (17,000) and Sir William Johnson in 1763 (10,000).

33. Marshall, Orsamus. "Narrative of the Expedition of Marquis de Nonville against the Senecas, in 1687." *Collections of the New York Historical Society*. Second Series, Vol. II, 1849: 155–6, 182.

34. Parker, 1910: 62–63.

35. Parker: 29–32 Roger Williams described the same communal work system among the New England Algonquins, "As an organized body of workers, the women of each gens (clans) formed a distinct agricultural corporation." (Stites in Parker, 1910: 30.)

36. Waugh. F.W. *Iroquois Foods and Food Preparation*. Ottawa, Canada: National Museum of Man, No. 12, Anthropological Series, 1916, Rpt. 1973: 11–12.

37 Parker, 1910: 21–22. Waugh: 9.

38. Lewandowski, Stephen. "Diohe'ko, The Three Sisters in Seneca Life: Implications for a Native Agriculture in the Finger Lakes Region of New York State." *Agriculture and Human Values*, 1987, Spring/Summer: 91–92.

39. Sauer, Carl O. *Agricultural Origins and Dispersals*. Cambridge: M.I.T. Press, 1952. Rpt. 1969: 64.

40. Waugh: 21, Lewandowski: 82.

41. Parker, 1910: 26.

42. Waugh: 18–20.

43. Parker, 1916: 54.

44. Mt. Pleasant, Jane. "The Iroquois Sustainers: Practices of a Longterm Agriculture in the Northeast." *Indian Corn of the Americas*. American Indian Program, Cornell University, Ithaca, N.Y. 1989: 33–39.

45. Parker, 1910: 27.

46. Lewandowski: 82. Parker, 1910: 29.

47. Parker, 1910: 29.

48. Waugh: 12–14.

49. Mt. Pleasant: 35–36.

50. Mt. Pleasant: 35–36. Quintana, Jorge. "Indian Corn Agriculture: A Puzzle for the 21st Century." *Indian Corn of the Americas*. American Indian Program, Cornell University, Ithaca, N.Y., 1989: 28–32.

51. Lewandowski: 83–84.

52. Mt. Pleasant: 37–38.

53. Parker, 1910: 43. The Mohawk varieties of corn, and the Mohawk names, were given by William Loft of Six Nations Reserve in Canada: Tuscarora, Tuscarora short, sweet corn, hominy or white flint, hominy long eared, yellow, purple, husk or pod, pop. (Waugh).

54. Lewandowski: 91.

55. Mt. Pleasant: 39.

56. In the 1840s, Morgan gathered numerous items made by the Haudenosaunee however most of Morgan's collection was lost in the fire of the state museum in 1911. Many examples of Haudenosaunee technology were gathered and extensive manuscripts were published which provide photographs and sketches of these items (Parker, 1910; Waugh, 1916; Lyford, 1945; Harrington, 1908). Vast numbers of photographs, sketches, and the actual items are readily available in the above mentioned sources and museums. See Parker: 1910: 32–55 for descriptions of tools.

57. Lafitau quoted by Parker, 1910: 34. A photograph caption reads: "Braid of Seneca calico-hominy corn. This is the native method of preserving dried corn on the cob, now widely adopted by white people and others."

58. Lewandowski: 82, Parker, 1910: 35, Waugh: 41–42.

59. Parker, 1910: 34–36, Waugh: 171–175.

60. Lewandowski: 84.

61. Parker,1910: 66–80. Waugh: 78–103.

62. Cornplanter: 18.

63. Harrington: 583–584.

64. Katz, S. H., M. L. Hediger, and L. A. Valleroy. "Traditional Maize Processing Techniques in the New World." *Science*, Vol. 184, 17 May 1974: 765–773. Lewandowski: 84.

65. Waugh: 90.

66. Parker, 1910: 69–76 Harrington: 587.

67. Parker, 1910: 67.

68. Parker, 1910: 66–78. Waugh, Harrington, Lyford.

69. Parker, 1909: 177.

70. Mohawk interview.

71. Lewandowski: 91.

Chapter 7   The Interaction of Corn and Cultures

1. Hurt, R. Douglas. *Indian Agriculture in America, Prehistory to the Present*. Kansas: University Press of Kansas, 1987: 7.

2. Columbus (1492), the Jesuits (1600s), Kalm (1750), Morgan (1850), and Parker (1920) each provide information on corn.

3. Lewandowski, Stephen. "Diohe'ko, The Three Sisters in Seneca Life: Implications for a Native Agriculture in the Finger Lakes Region of New York State." *Agriculture and Human Values*, 1987, Spring/Summer: 78.

4. Sturtevant, E. Lewis. "Indian Corn and the Indian." *The American Naturalist*. Philadelphia: McCalla and Stavely, 1885: 227.

5. Parker, Arthur, C. *Iroquois Uses of Maize and Other Food Plants*. Albany, University of the State of New York: New York State Museum bulletin, No. 144, 1910: 17.

6. Sturtevant: 229.

7. Gehring, Charles, and William Starna. *A Journey into Mohawk and Oneida Country, 1634–1635*. Syracuse: Syracuse University Press, 1988.

8. Sturtevant: 229–30.

9. Hurt: 31.

10. Usner, Daniel, H. *Indians, Settlers, and Slaves in a Frontier Exchange Economy, The Lower Mississippi Valley Before 1783*. Chapel Hill: University of North Carolina Press, 1992: 170–171.

11. Marriott, Alice, and Carol K. Rachlin. *American Indian Mythology*. New York: Thomas Y. Crowell, 1968: 105–111.

12. Stuart, George E. "Who Were the 'Mound Builders'?" *National Geographic*. December 1972, Vol. 142, No. 6: 789–795.

13. Hurt: 35–36.

14. Plumb, Charles S. *Indian Corn Culture*. Chicago: Breeder's Gazette Print, 1895: 10.

15. Sturtevant: 230–1.

16. Sturtevant: 230–1.

17. Ortiz, Alfonso. "Some Cultural Meanings of Corn in Aboriginal North America," *Indian Corn of the Americas*, ed. Jose Barreiro, 1989: 70–71.

18. Hurt: 57–58.

19. Wilson, Gilbert L. *Buffalo Bird Woman's Garden*. St. Paul: Minnesota Historical Society Press, 1917. Rpt. 1987: 27.

20. Hurt: 42–46.

21. Hurt: 53–54.

22. Ortiz, Alfonso. *The Tewa World.* Chicago: University of Chicago Press,1969: 13– 14,30–32.

23. Ortiz, 1989: 65.

24. Ortiz, 1969: 51–68.

25. Ortiz, 1969: 103.

26. Ortiz, 1989: 69–73.

27. Sturtevant: 226.

28. Hurt: 3–4.

29. Teni, Roderico. "Cosmological Importance of Corn among Mayan Peoples." *Indian Corn of the Americas,* ed. Jose Barreiro: 14–18.

30. Sturtevant: 226–7.

31. Emerson, William D. *Indian Corn, and Its Culture.* Cincinnati, Ohio: Wrightson & Co., 1878: 24–27.

32. Sauer, Carl O. *Agricultural Origins and Dispersals.* Cambridge: M.I.T. Press, 1952. Rpt. 1969: 490–494.

33. Katz, S. H., M. L. Hediger, and L. A. Valleroy. "Traditional Maize Processing Techniques in the New World." *Science.* Vol. 184, 17 May 1974: 766–7.

34. Lewandowski, Stephen. "Diohe'ko, The Three Sisters in Seneca Life: Implications for a Native Agriculture in the Finger Lakes Region of New York State." *Agriculture and Human Values,* 1987, Spring/Summer: 80.

35. Warman, Arturo. "Maize as Organizing Principle: How Corn Shaped Space, Time, and Relationships in the New World." *Indian Corn of the Americas,* ed. Jose Barreiro: 20.

36. Barreiro, Jose. "View From the Shore, American Indian Perspectives on the Quincentenary." *Northeast Indian Quarterly.* Ithaca NY, Fall 1990: 68.

37. Barreiro, 1990: 71.

38. Barreiro, 1990: 69.

39. Warman: 21.

40. Weatherford, Jack. *Indian Givers—How the Indians of The Americas Transformed the World.* New York: Crown Publishers, 1988: 73.

41. Miracle, Marvin P. *Maize in Tropical Africa.* Madison: University of Wisconsin Press, 1966: 87–97.

42. Warman: 23–24.

43. Emerson: 28.

44. Bradford in Parker, Arthur C. *Iroquois Uses of Maize and Other Food Plants.* Albany, University of the State of New York: New York State Museum bulletin, No.144, 1910: 14.

45. See: Bradford, *History of the Plymouth Plantation.* Cols. Mass. Hist. Soc. Boston, 1856, Ser. 4, III: 130. Ceci, Lynn. "Squanto and the Pilgrims." *Society,* May/June, 1990: 41. Josephy, Alvin M. *The Indian Heritage of America.* New York: Alfred A. Knopf, 1968: 301. Hurt: 39. Hoxie, Fedrick E. *"The Indian Versus the Textbooks: Is There Anyway Out?"* Newberry Library, 1984: 20.

46. Hurt: 38–39.

47. Ceci: 41–44.

48. Parker, 1910: 14–15.

49. Emerson: 31–33.

50. Lawson, John. *Lawson's History of North Carolina.* Richmond, Virginia: Garrett and Massie, 1714: 76.

51. Wessel, Thomas R. "Agriculture, Indians, and American History." *Agricultural History,* 50 (January 1976): 3–4.

52. Kalm, Pehr. "Description of Maize, How It Is Planted and Cultivated in North America, Together with the Many Uses of this Crop Plant." *Agricultural History,* IX, April, 1935: 103–114.

53. Wessel: 5–6.

54. Crosby, Alfred W. "Maize, Land, Demographics, and the American Character." *Revue Francaise D'Etudes Americaines,* April-June 1991: 152–156.

55. Marshall, Orsamus. "Narrative of the Expedition of Marquis de Nonville against the Senecas, in 1687." *Collections of the New York Historical Society.* Second Series, Vol. II, 1849: 10.

56. Marshall: 165.

57. O'Callaghan. E. B. *The Documentary History of the State of New York.* Vol. I, Albany, New York: Weed, Parson & Co. 1849: 238–239.

58. Marshall: 152.

59. Marshall: 152–160.

60. Mohawk, John. "Economic Motivations—An Iroquoian Perspective." In *Indian Corn of the Americas, Gift to the World.* ed. Jose Barreiro. Northeast Indian Quarterly, Ithaca, New York Spring/Summer, 1989: 11.

61. Marshall: 182. O'Callaghan: 239.

62. Bancroft, George. *History of the United States of America,* Boston, Massachusetts: Little, Brown and Co. 1878.: 79.

63. Plumb: 9–10.

64. Accounts of the battles can be found in Sturtevant (1885), Cook (1887), Graymont (1972), and Abler (1989).

65. Abler, Thomas S. *Chainbreaker: The Revolutionary War Memoirs of Governor Blacksnake, As told to Benjamin Williams.* (1753–1859) Lincoln: University of Nebraska Press, 1989: 36, 230–231.

66. Parker, Arthur C. *The History of the Seneca Indians.* Empire State Historical publication, 1926, Rpt. Ira J. Friedman, Inc. 1967: 122.

67. Parker, 1926: 124.

68. Abler: 273.

69. Abler: 113.

70. Cook, Frederick. *Journals of the Military Expedition of Major General John Sullivan Against the Six Nations of Indians in 1779, with Records of Centennial Celebrations.* Auburn: Knapp, Peck & Thomson, 1887: 5.

71. Parker, 1926: 125.

72. Conover's Early History of Geneva, N.Y: 47. Sturtevant: 228.

73. Cook: 23.
74. Cook: 27.
75. Cook: 48.
76. Cook: 26–28, 46–47.
77. Parker, 1910: 20.
78. Stone, William. *Life of Joseph Brant—Thayendanegea, Including the Indian Wars of the American Revolution.* New York: George Dearborn & Co., 1838: 33.
79. Cook: 305.
80. Rothenberg, Diane. "The Mothers of the Nation: Seneca Resistance to Quaker Intervention." In Mona Etienne, and Eleanore Leacock, *Women and Colonization.* New York: Praeger Publishers, 1980: 64.
81. Wallace, Henry A., and William L. Brown. *Corn and Its Early Fathers.* Michigan State University Press, 1956.: 221–222.
82. Wallace: 223–226,274–302.
83. Rothenberg: 77–78.
84. Mohawk interview.
85. Parker, 1910.
86. Brown, Judith K. "Economic Organization and the Position of Women Among the Iroquois." *Ethnohistory* 17, 3–4, 1970: 159.
87. Wallace: 275.
88. Wallace: 282.
89. Giles, Dorothy. *Singing Valleys, the Story of Corn.* New York: Random House, 1940: 105.
90. Hardeman, Nicholas P. *Shucks, Shocks, and Hominy Blocks, Corn As a Way of Life in Pioneer America.* Baton Rouge: Louisiana State University Press, 1981: 27.
91. Giles: 104.
92. Hardeman: 77.
93. Enfield, Edward. *Indian Corn: Its Value, Culture, and Uses.* New York: D. Appleton and Company, 1866: 92, 127–131.
94. Plumb, 1895: 88.
95. Hardeman: 5, 34–38.
96. Giles: 212.
97. Murphy, Charles J. *American Indian Corn: 150 Ways to Prepare and Cook It.* New York: G.P. Putnam's Sons, 1917: iii, v–vi, x–xi.
98. Enfield: 60–66, 71–75. Plumb: 20–40.
99. Wallace: 16.
100. Enfield: 12–20.
101. Furnas, Robert W. *Corn: Its Origin, History, Uses and Abuses.* Lincoln, Nebraska: Journal Company, State Printers, 1886: 5–17.
102. Murphy: 8–13.
103. Warman: 27.
104. Quintana, Jorge. "Indian Corn Agriculture: A Puzzle for the 21st Century." *Indian Corn of the Americas*, ed. Jose Barreiro. American Indian Program, Cornell University, Ithaca, N.Y., 1989: 29.
105. Enfield: 234–236.

106. Emerson: 206–215.
107. Murphy: 47.
108. Wallace and Brown: 20. Enfield: 29. Hardeman: 153.
109. Hardeman: 151.
110. Emerson: 229–320.
111. Emerson: 232. Hardeman: 149.
112. Enfield: 282.
113. Murphy: 21. Emerson: 233. Enfield: 279. Hardeman: 149.
114. Elting, Mary, and Michael Folsom. *The Mysterious Grain, Science in Search of the Origin of Corn.* New York, New York: M. Evans and Company, 1967: 106.
115. Warman: 26–7.
116. Wallace and Brown: 19.
117. Hardeman: 231.
118. Hardeman: 5.

Chapter 8   Dynamic Aspects of Haudenosaunee Culture

1. Gonyea, Ray. Introduction by Fred R. Wolcott. *Onondaga: Portrait of a Native People.* Syracuse: Syracuse University Press, 1968: 11, 22.

2. Hauptman in Wolcott, *Onondaga: Portrait of a Native People.* Syracuse: Syracuse University Press, 1968: 9.

3. *American Indian Treasures* brochure, Rochester Museum, Rochester, N.Y.

4. *American Indian Treasures.*

5. Hauptman, Laurence, M. "The Iroquois Schools of Art: Arthur C. Parker and the Seneca Arts Project, 1935–1941." *New York History*, July 1979: 293.

6. Fenton, William. "Aboriginally Yours, Jesse J. Cornplanter, Hahyonh-wonh-Ish, The Snipe." Ed. Margot Liberty. *American Indian Intellectuals.* St. Paul, Minnesota: West Publishing, 19781978: 180. Fenton says that Jesse became tired of being the "informant" and wanted to become an author. On a typewriter he borrowed from Fenton, Jesse wrote *Legends of the Longhouse* which was published in 1938. (Fenton, 1978).

7. Hauptman: 286.

8. Hauptman: 288. Hertzberg, Hazel. "Arthur C. Parker." Ed. Margot Liberty, *American Indian Intellectuals.* St. Paul, Minnesota: West Publishing, 1978: 128–138.

9. Parker, Arthur C. *The Life of General Ely S. Parker.* Buffalo: Buffalo Historical Society, 1919: 565–56, 69–72.

10. Fenton in Parker, Introduction, *Parker on the Iroquois.* New York: Syracuse University Press, 1968: 1.

11. Parker, 1919: 79.

12. Tooker, Elisabeth. "Ely S. Parker." Ed. Margot Liberty, *American Indian Intellectuals.* St. Paul, Minnesota: West Publishing, 1978: 15–30: 16–19.

13. *Images from the Longhouse, Paintings of Iroquois Life by Seneca Artist Ernest Smith.* Rochester, New York: Rochester Museum and Science Center, brochure, 1975.

14. Parker, Arthur. *Life and Legends of the Iroquois Indians.* Rochester: Rochester Museum, pamphlet, 1939.

15. Rochester Museum sign.

16. Rochester Museum sign.

17. *Iroquois Trail.* New York, Rochester Museum and Science Center, brochure.

18. Hayes, Charles F. "Ernest Smith, Arthur C. Parker, and the Rochester Museum." In *Iroquois Studies: A Guide to Documentary and Ethnographic Resources from Western New York and the Genesee Valley,* ed. Russell A. Judkins. Dept. of Anthropology, SUNY and the Geneseo Foundation, 1987: 17–19.

19. Parker, 1939.

20. Tehanetorens. *Tales of the Iroquois.* Rooseveltown, New York: Akwesasne Notes, 1976: 69–74.

21. Hewitt, J. N. B. *Iroquoian Cosmology.* Annual Reports Bureau of Ethnology, Part 1, 21st report 1899–1900, Part 2, 43rd report, 1925–1926: 223–224.

22. Parker, Arthur, C. *The Code of Handsome Lake, the Seneca Prophet.* Albany, University of the State of New York, 1913: 54.

23. Parker, Arthur, C., *Iroquois Uses of Maize and Other Food Plants.* Albany, University of the State of New York: New York State Museum bulletin, No. 144, 1910: 40.

24. *Iroquois Trail* brochure, Rochester Museum.

25. By Erminnie Smith (1883), Hewitt (1899), Curtin and Hewitt (1910), Arthur Parker (1923), Jesse Cornplanter (1938), and Tehanetorens (1976).

26. See appendix for list of stories not included in this chapter which can be found in the above mentioned publications.

27. Fenton, "Aboriginally Yours, Jesse J. Cornplanter, Hah-yonh-wonh-Ish, The Snipe." *American Indian Intellectuals,* ed. Margot Liberty. St. Paul, Minnesota: West Publishing, 1978: 176–195. Hertzberg, Hazel. "Arthur C. Parker." *American Indian Intellectuals*, ed. Margot Liberty. St. Paul, Minnesota: West Publishing, 1978: 128–138. Tooker.

28. Wolcott, Fred R. *Onondaga, Portrait of a Native People.* Syracuse: Syracuse University Press, 1986: 3.

29. Fenton, William. "Tonawanda Longhouse Ceremonies: Ninety Years After Lewis Henry Morgan." *Bureau of American Ethnology,* Bulletin 128, Smithsonian Institution, 1941: 138–165.1941.

30. Wall, Steve, and Harvey Arden. *WisdomKeepers.* Hillsboro, Oregon: Beyond Words Publishing, Inc., 1990.

31. Caduto, Michael J., and Joseph Bruchac. *Keepers of the Earth, Native American Stories and Environmental Activities for Children.* Golden, Colorado: Fulcrum, Inc., 1988.

32. Cole, Donna. "Cornhusk People." *Unbroken Circles,* ed. Susan Dixon. Ithaca, New York: *Northeast Indian Quarterly*, Winter, 1990: 62–64.

33. Cole: 64.

34. Arden, Harvey. "The Fire That Never Dies," *National Geographic.* September, 1987: 375–403.

Chapter 9   The Contemporary Role of Corn in Haudenosaunee Culture

1. Sosa, John. *The Maya Sky, The Maya World: A Symbolic Analysis of Yucatec Maya Cosmology.* Diss. SUNY-Albany, 1985: 116. Ortiz, Alfonso. *New Perspectives on the Pueblos.* Albuquerque: University of New Mexico Press, 1972.

2. Austin, Alberta. *Ne'Ho Niyo' De:no', That's What It Was Like.* Irving, New York: Seneca Nation Curriculum Development Project, 1986. Graymont, Barbara. *The Iroquois in the American Revolution.* Syracuse: Syracuse University Press, 1972. Mohawk, John interview.

3. USDA, Census Bureau.

4. Quintana, Jorge. "Agricultural Survey of New York State Iroquois Reservations, 1990." Ithaca, New York. *Northeast Indian Quarterly*, Spring, 1991: 34–35.

5. Mohawk interview.

6. Powless, Irv Sr. interview.

7. Cook, Katsi interview.

8. *Iroquois Cookbook.* Peter Doctor Memorial Indian Scholarship Foundation, Allegany Reservation, 1989.

9. Barreiro, Jose interview.

10. Barreiro, Jose. "First Words" *Northeast Indian Quarterly*, Summer 1991: 2–3.

11. Quintana, 1991: 36.

12. Mt. Pleasant, Jane. "The Iroquois Sustainers: Practices of a Longterm Agriculture in the Northeast." *Indian Corn of the Americas,* ed. Jose Barreiro. American Indian Program, Cornell University, Ithaca, N.Y., 1989: 33–39.

13. Cook, Katsi interview.

14. Barreiro interview: 5–6.

15. Parker, Arthur, C., *Iroquois Uses of Maize and Other Food Plants.* Albany, University of the State of New York: New York State Museum bulletin, No. 144, 1910: 43.

16. Cutler, 293.

17. Elting, Mary, and Michael Folsom. *The Mysterious Grain, Science in Search of the Origin of Corn.* New York, New York: M. Evans and Company, 1967: 81.

18. Mangelsdorf, P. C., and R. G. Reeves. *The Origin of Indian Corn and Its Relatives* Texas Agricultural Experiment Station, Bulletin 574, May 1939: 220.

19. Mangelsdorf: 217.

20. Mangelsdorf: 46.

Bibliography

Abler, Thomas S. *Chainbreaker: The Revolutionary War Memoirs of Governor Blacksnake, As told to Benjamin Williams* (1753–1859). Lincoln: University of Nebraska Press, 1989.

Ahlquist, Roberta. "Developing Our Diverse Voices: Critical Pedagogy for a Multicultural Classroom." Paper presented at *American Educational Research Association Meeting*. San Francisco, California, March 1989.

Allport, Gordon W. *The Nature of Prejudice*. Garden City, New York: Addison-Wesley Publishing, 1958.

Arden, Harvey. "The Fire That Never Dies." *National Geographic* (September 1987): 375–403.

Austin, Alberta. *Ne'Ho Niyo' De:no', That's What It Was Like*. Irving, New York: Seneca Nation Curriculum Development Project, 1986.

———. *Ne'Ho Niyo' De: No', That's What It Was Like, Volume II*. Irving, New York: Seneca Nation Curriculum Development Project, 1989.

Bancroft, George. *History of the United States of America*. Boston, Massachusetts: Little, Brown and Co., 1878.

Banks, James A., and James Lynch. *Multicultural Education in Western Societies*. Westport, Connecticut: Praeger, 1986.

———. *Multicultural Education, Issues and Perspectives*. Needham Height, Massachusetts: Allyn and Bacon, 1989.

Barco, Mandalit del. *Hispanic* (September 1990): 15–18.

Barreiro, Jose. *Indian Corn of the Americas, Gift to the World*. Ithaca, New York: *Northeast Indian Quarterly*, Spring/Summer 1989.

———. *View From the Shore, American Indian Perspectives on the Quincentenary*. Ithaca, NY, Fall 1990.

———. "First Words." *Northeast Indian Quarterly* (Summer 1991): 2–3.

———. *Native Corn Report*. Cornell American Indian Agriculture Program. Ithaca, New York: *Northeast Indian Quarterly*, Summer 1991.

Beauchamp, William, M. *Civil, Religious and Mourning Councils and Ceremonies of Adoption of the New York Indians*. New York State Museum Bulletin 113, 1907.

Beck, Peggy V., and Anna L. Walters. *The Sacred Ways of Knowledge, Source of Life*. Tsaile, Arizona: Navajo Community College, 1977.

Berkhofer, Robert F. *The White Man's Indian: Images of the American Indian from Columbus to the Present*. New York: Random House, 1978.

Bradford. *History of the Plymouth Plantation Cols*. Mass. Hist. Soc. Boston 1856, Ser. 4, III.

Brown, Judith K. "Economic Organization and the Position of Women Among the Iroquois." *Ethnohistory* 17, 3–4 (1970): 151–167.

Butterfield, Robin A. "The Development and Use of Culturally Appropriate Curriculum for American Indian Students." *Peabody Journal of Education*—The Transcultural Education of American Indian and Alaska Native Children: 49–66.

Caduto, Michael J., and Joseph Bruchac. *Keepers of the Earth, Native American Stories and Environmental Activities for Children*. Golden, Colorado: Fulcrum, Inc., 1988.

Campisi, Jack, and Laurence Hauptman. *The Oneida Indian Experience, Two Perspectives*. New York: Syracuse University Press, 1988.

Ceci, Lynn. "Squanto and the Pilgrims." *Society* (May/June 1990): 40–44.

Chafe, Wallace. *Seneca Thanksgiving Rituals*. Bureau of American Ethnology, Bulletin 183, Smithsonian Institution. Washington, DC: U.S. Government Printing Office, 1961.

Charles, James P. "The Need for Textbook Reform: An American Indian Example." *Journal of American Indian Education* (May 1989): 1–13.

Charles, Ramona. "Tonawanda Seneca Indian Reservation." *A Genesee Harvest*:86–87.

Cole, Donna. "Cornhusk People." In *Unbroken Circles*, ed. Susan Dixon. Ithaca, New York: *Northeast Indian Quarterly*, Winter 1990:62–64.

Converse, Harriet Maxwell. *Myths and Legends of the New York State Iroquois*, ed. Arthur Parker. New York State Museum Bulletin 125, 1908.

Cook, Frederick. *Journals of the Military Expedition of Major General John Sullivan Against the Six Nations of Indians in 1779, with Records of Centennial Celebrations.* Auburn: Knapp, Peck & Thomson, 1887.

Cornplanter, Jesse J. *Legends of the Longhouse.* Philadelphia, Pennsylvania: J.B. Lippincott, 1938.

Council on Interracial Books for Children. *Stereotypes, Distortions, and Omissions in U.S. History Textbooks.* New York, 1977.

Crosby, Alfred W. "Maize, Land, Demographics, and the American Character." *Revue Francaise D'Etudes Americaines* (April–June 1991): 151–162.

Curriculum of Inclusion. Report of the Commissioner's Task Force on Minorities: Equity and Excellence, New York State Education Department, July, 1989.

Curtin, Jeremiah, and J. N. B. Hewitt. *Seneca Fictions, Legends, and Myths.* 32nd Annual Report of the Bureau of American Ethnology, Smithsonian Institution, 1910–11.

DeLabarre, Edmund B., and Harris H. Wilder. "Indian Corn-Hills in Massachusetts." *American Anthropologist* 22: 203–225 (July–September 1920).

Dixon, Susan. *Unbroken Circles, Traditional Arts of Contemporary Woodland Peoples.* Ithaca, New York: *Northeast Indian Quarterly*, Winter 1990.

Ellison, Ralph. *Going to the Territory.* New York: Random House, 1986.

Elson, Ruth. *Guardian of Traditions: American Schoolbooks of the Nineteenth Century.* Lincoln: University of Nebraska Press, 1964, 71–81.

Elting, Mary, and Michael Folsom. *The Mysterious Grain, Science in Search of the Origin of Corn.* New York, New York: M. Evans and Company, 1967.

Emerson, William D. *Indian Corn, and Its Culture.* Cincinnati, Ohio: Wrightson & Co., 1878.

Enfield, Edward. *Indian Corn; Its Value, Culture, and Uses.* New York: D. Appleton and Company, 1866.

Ernest Smith, Iroquois Artist. American Indian Treasures, Amsterdam, New York: Noteworthy Co., brochure.

Faces. "The Iroquois." Peterborough, New Hampshire: Cobblestone Publishing, September 1990.

Fenton, William. "This Island, the World on the Turtle's Back." *Journal of American Folklore* 75 (1962): 283–300.

———. "An Outline of Seneca Ceremonies at Coldspring Longhouse." *Yale University Publications in Anthropology, Number 9.* Yale University Press, 1936: 3–23.

———. "Tonawanda Longhouse Ceremonies: Ninety Years After Lewis Henry Morgan." Bureau of American Ethnology, Bulletin 128, Smithsonian Institution, 1941: 138–165.

———. "Tonawanda Reservation 1935: The Way It Was." In *The Iroquois In the American Revolution, 1976 Conference Proceedings, Research Records No. 14,* ed. Charles Hayes. Rochester, New York: Rochester Museum, 1981.

———. Introduction, *Parker on the Iroquois.* Syracuse: Syracuse University Press, 1968.

———. "Aboriginally Yours, Jesse J. Cornplanter, Hah-yonh-wonh-Ish, The Snipe." In *American Indian Intellectuals,* ed. Margot Liberty. St. Paul, Minnesota: West Publishing, 1978, 176–195.

Ferguson, Maxel J., and Dan B. Fleming. "Native Americans in Elementary School Social Studies Textbooks." *Journal of American Indian Education* 23 (2) (January 1984): 10–15.

FitzGerald, Frances. *America Revised—History of Schoolbooks in the Twentieth Century.* Boston: Little Brown & Co., 1979.

Foster, Micheal F. *From the Earth to Beyond the Sky: An Ethnographic Approach to Four Longhouse Iroquois Speech Events.* Ottawa: National Museums of Canada, 1974.

Fuchs, Estelle, and Robert J. Havighurst. *To Live on This Earth, American Indian Education.* Garden City: Doubleday, 1972.

Furnas, Robert W. *Corn: Its Origin, History, Uses and Abuses.* Lincoln, Nebraska: Journal Company, State Printers, 1886.

Gagnon, Paul. *Democracy's Untold Story, What World History Textbooks Neglect.* American Federation of Teachers, 1987.

Garcia, Jesus. "From Bloody Savages to Heroic Chiefs." *Journal of American Indian Education* 17(2). Tempe, Arizona: Arizona State University: 1978: 15–19.

Gay, Geneva, and James A. Banks. "Teaching the American Revolution: A Multiethnic Approach." *Social Education* 39 (November–December 1975): 461–465.

Geertz, Clifford. *The Interpretation of Cultures.* New York: Basic Books, 1973.

———. "Ethos, World-View and the Analysis of Sacred Symbols." *Antioch Review*, 17(1957): 421–437.

Gehring, Charles, and William Starna. *A Journey into Mohawk and Oneida Country, 1634–1635.* Syracuse: Syracuse University Press, 1988.

Genesee Valley Council on the Arts. *A Genesee Harvest, A Scene in Time, 1779.* Rochester, New York: Early Arts in the Genesee Valley Series: No. 7, 1979.

Giles, Dorothy. *Singing Valleys, the Story of Corn.* New York: Random House, 1940.

Gonyea, Ray. Introduction. In Fred R. Wolcott. *Onondaga: Portrait of a Native People.* Syracuse: Syracuse University Press, 1968, 11–32.

Goodenough, Ward H. *Culture, Language, and Society.* Addison-Wesley Module in Anthropology, 1971.

Graymont, Barbara. *The Iroquois in the American Revolution.* Syracuse: Syracuse University Press, 1972.

———. "The Six Nations Indians in the Revolutionary War," In *The Iroquois in the American Revolution, 1976 Conference Proceedings, Research Records No. 14,* ed. Charles Hayes. Rochester, New York: Rochester Museum, 1981.

———. *Fighting Tuscarora, the Autobiography of Chief Clinton Rickard.* Syracuse: Syracuse University Press, 1973.

Hallowell, A. I. "Indian Corn Hills." *American Anthropologist,* N.S. 23 (1921): 233.

Hammell, George "The Prehistory of the Seneca Iroquois in the Genesee Valley." *A Genesee Harvest,* 19–36.

———. "The Genesee Valley Seneca Iroquois, 1750–97: Contemporary Sources." *A Genesee Harvest,* 61–85.

Hardeman, Nicholas P. *Shucks, Shocks, and Hominy Blocks, Corn As a Way of Life in Pioneer America.* Baton Rouge: Louisiana State University Press, 1981.

Harrington, M. R. "Some Seneca Corn-Foods and Their Preparation." *American Anthropologist* N. S. 10 (1908): 575–590.

Harris, Helen L. "On the Failure of Indian Education." *The Clearing House* 48 (December 1973): 242–247.

Hauptman, Laurence M. "The Iroquois Schools of Art: Arthur C. Parker and the Seneca Arts Project, 1935–1941." *New York History* (July 1979): 282–312.

Hayes, Charles F. "Ernest Smith, Arthur C. Parker, and the Rochester Museum." In *Iroquois Studies: A Guide to Documentary and Ethnographic Resources from Western New York and the Genesee Valley,* ed. Russell A. Judkins. Dept. of Anthropology, SUNY and the Geneseo Foundation, 1987: 17–19.

Henry, Jeannette, and Rupert Costo. *Textbooks and the American Indian.* California: American Indian Historical Society, 1970.

Hertzberg, Hazel. "Arthur C. Parker." In *American Indian Intellectuals,* ed. Margot Liberty. St. Paul, Minnesota: West Publishing, 1978: 128–138.

Hewitt, J. N. B. *Iroquoian Cosmology.* Annual Reports Bureau of Ethnology, Part 1, 21st report 1899–1900, Part 2, 43rd report, 1925–1926.

Hirschfelder, Arlene B. *American Indian Stereotypes in the World of Children.* Metuchen, New Jersey: Scarecrow Press, 1982.

———. "The Treatment of Iroquois Indians in Selected American History Textbooks." *The Indian Historian* 2(1975): 31–39.

Hobson, Henry. *Seeds of Change.* New York: Harper and Row, 1986.

Horne, Gerald. *Thinking and Rethinking U.S. History.* Council on Interracial Books for Children, 1988.

Hoxie, Fedrick E. *"The Indian Versus the Textbooks: Is There Anyway Out?"* Newberry Library, 1984.

Hurt, R. Douglas. *Indian Agriculture in America, Prehistory to the Present.* Kansas: University Press of Kansas, 1987.

Images From the Longhouse, Paintings of Iroquois Life by Seneca Artist Ernest Smith. Rochester: Rochester Museum and Science Center, brochure, 1975.

Indian Education: A National Tragedy—A National Challenge. Washington, D.C.: U.S. Government Printing Office, 1969.

Indian Nations At Risk: An Educational Strategy for Action. United States Department of Education, Final Report of the Indian Nations At Risk Task Force, 1991.

Iroquois Cookbook. Peter Doctor Memorial Indian Scholarship Foundation, Allegany Reservation, 1989.

Iroquois Recipes, Salamanca, New York: Seneca Bilingual Education Program, brochure.

Iroquois Trail. New York: Rochester Museum and Science Center, brochure.

Jenness, Diamond. *The Corn Goddess.* Ottawa. Canada: National Parks Branch, National Museum of Canada, Bulletin no. 141, Anthropological Series no. 39, 1956.

Josephy, Alvin M. *The Indian Heritage of America.* New York: Alfred A. Knopf, 1968.

Joyce, Bruce, and Marsha Weil. *Models of Teaching.* Englewood Cliffs, New Jersey: Prentice-Hall, 1986.

Kalm, Pehr. "Description of Maize, How it Is Planted and Cultivated in North America, Together with the Many Uses of this Crop Plant." *Agricultural History* IX (April 1935): 98–117.

Katz, S. H., M. L. Hediger, and L. A. Valleroy. "Traditional Maize Processing Techniques in the New World." *Science* 184 (17 May 1974): 765–773.

Kipp, Henry W. *Indians in Agriculture: A Historical Sketch.* U.S. Department of the Interior, Bureau of Indian Affairs, 1988.

Kliebard, Herbert. *The Struggle for the American Curriculum 1893–1958.* New York: Routledge & Kegan Paul, 1987.

Kuper, Adam. *The Invention of Primitive Society.* New York: Routledge, 1988.

Lanman, Charles. "Green-Corn Ceremonies of the Cherokees." *The Magazine of History with Notes and Queries.* Poughkeepsie, New York: Tarrytown, W. Abbatt, Vol. 19, August–September 1914: 89–92.

Lawson, John. *Lawson's History of North Carolina.* Richmond, Virginia: Garrett and Massie, 1714.

Lewandowski, Stephen. "Diohe'ko, The Three Sisters in Seneca Life: Implications for a Native Agriculture in the Finger Lakes Region of New York State." *Agriculture and Human Values* (1987, Spring/Summer) 76–93.

———. "Three Sisters—An Iroquoian Cultural Complex." *Indian Corn of the Americas, Gift to the World.* Ithaca, New York: *Northeast Indian Quarterly*, Spring/Summer 1989, 41–45.

Lincoln, Yvonna S., and Egon G. Guba. *Naturalistic Inquiry.* Newbury Park, California: Sage Publications, 1985.

Lyford, Carrie. *Iroquois, Their Art and Crafts.* 1945. rpt. as *Iroquois Crafts.* Blaine, Washington: Hancock House Publishers, 1989.

Malefijut de Waal, Annemarie. *Images of Man, A History of Anthropological Thought.* New York: Alfred A. Knopf, 1974.

Mangelsdorf, P. C., and R. G. Reeves. *The Origin of Indian Corn and Its Relatives.* Texas Agricultural Experiment Station, Bulletin 574, May 1939.

Marriott, Alice, and Carol K. Rachlin. *American Indian Mythology.* New York: Thomas Y. Crowell, 1968.

Marshall, Orsamus. "Narrative of the Expedition of Marquis de Nonville against the Senecas in 1687." *Collections of the New York Historical Society.* Second Series, Vol. II, 1849, 149–192.

Milwaukee Public Schools. *K–12 Curriculum, Report of the K–12 Curriculum Committee.* January 1990.

Miracle, Marvin P. *Maize in Tropical Africa*. Madison: University of Wisconsin Press, 1966.

Mohawk, John. "Economic Motivations—An Iroquoian Perspective." Ed. Jose Barreiro. *Indian Corn of the Americas, Gift to the World*. Ithaca, New York: Northeast Indian Quarterly, Spring/Summer, 1989: 56–63.

———. *War Against the Seneca, The French Expedition of 1687*. Victor, New York: Ganondagan State Historic Site.

Morgan, Lewis Henry. *League of the Ho-De-No-Sau-Nee or Iroquois*. New York: Burt Franklin, 1850. Rpt. 1901.

———. *Houses and House-Life of the American Aborigines*. Chicago: University of Chicago Press, 1881. Rpt. 1965.

Morris, Nancy Tucker. *A Method for Developing Multicultural Education Programs: An Example Using Cahuilla Indians Oral Narratives*. Diss. Iowa State University, 1976.

Mt. Pleasant, Jane. "The Iroquois Sustainers: Practices of a Longterm Agriculture in the Northeast." In *Indian Corn of the Americas,* ed. Jose Barreiro, 33–39.

Murphy, Charles J. *American Indian Corn: 150 Ways to Prepare and Cook It*. New York: G.P. Putnam's Sons, 1917.

O'Callaghan, E. B. *The Documentary History of the State of New York*. Vol. I, Albany, New York: Weed, Parson & Co., 1849.

Ogawa, Rodney T., and Betty Malen. "Towards Rigor in Reviews of Multivocal Literatures: Applying the Exploratory Case Study Method." *Review of Educational Research* 6 (Fall 1991): 225–286.

Oneida Nation Bilingual/Bicultural Program. *We Give Thanks to the Creator*. Wisconsin: Oneida Nation.

Onion, Daniel K. "Corn in the Culture of the Mohawk Iroquois." *Economic Botany* 18 (January–March 1964): 60–66.

One Nation, Many Peoples: A Declaration of Cultural Interdependence. Report of the New York State Social Studies Review and Development Committee, June, 1991.

O'Neill, G. Patrick. "The North American Indian in Contemporary History and Social Studies Textbooks." *Journal of American Indian Education* (May, 1987): 22–28.

Ortiz, Alfonso. "Some Cultural Meanings of Corn in Aboriginal North America." *Indian Corn of the Americas*. 1989, ed. Jose Barreiro. 64–73.

———. *The Tewa World*. Chicago: University of Chicago Press, 1969.

———. *New Perspectives on the Pueblos*. Albuquerque: University of New Mexico Press, 1972.

Parker, Arthur, C. *Iroquois Uses of Maize and Other Food Plants.* Albany, University of the State of New York: New York State Museum bulletin, No. 144, 1910.

———. *The Code of Handsome Lake, the Seneca Prophet.* Albany, University of the State of New York, 1913.

———. *The Constitution of the Five Nations.* New York State Museum Bulletin No. 184, 1916.

———. *Parker on the Iroquois,* edited with an introduction by William N. Fenton. (Includes the above in one volume.) Syracuse: Syracuse University Press, 1968.

———. *Seneca Myths and Folktales.* Buffalo: Historical Society, 1923. Rpt. University of Nebraska Press, 1989.

———. *The History of the Seneca Indians.* Empire State Historical publication, 1926. Rpt. Ira J. Friedman, Inc., 1967.

———. "Secret Medicine Societies of the Seneca." *American Anthropologist,* New. Ser. 1909, V. 2, No. 2: 161–185.

———. *Life and Legends of the Iroquois Indians.* Rochester: Rochester Museum, pamphlet, 1939.

———. *The Life of General Ely S. Parker.* Buffalo: Buffalo Historical Society, 1919.

Pearce, Roy Harvey. *Savagism and Civilization.* Baltimore: Johns Hopkins Press, 1953.

Plumb, Charles S. *Indian Corn Culture.* Chicago: Breeder's Gazette Print, 1895.

Posner, George, and Alan N. Rudnitsky. *Course Design, A Guide to Curriculum Development for Teachers.* White Plains, New York: Longman Inc. 1986.

———. *Analyzing the Curriculum.* New York: McGraw Hill, Inc., 1992.

Quintana, Jorge. "Indian Corn Agriculture: A Puzzle for the 21st Century." In *Indian Corn of the Americas,* ed. Jose Barreiro, 28–32.

———. "Agricultural Survey of New York State Iroquois Reservations, 1990." Ithaca, New York. *Northeast Indian Quarterly,* Spring 1991: 32–36.

Ravitch, Diane, and Chester E. Finn Jr. *What Do Our 17-Year-Olds Know.* New York: Harper and Row, 1987.

———. "Multiculturalism." *The American Scholar* (Summer 1990): 337–354.

———. "Remaking New York's History Curriculum." *New York Times,* Sept. 12, 1990.

Rothenberg, Diane. "The Mothers of the Nation: Seneca Resistance to Quaker Intervention." In Mona Etienne and Eleanore Leacock, *Women and Colonization*. New York: Praeger Publishers, 1980.

Rouse, Linda P., and Jeffery R. Hanson. "American Indian Stereotyping, Resource Competition, and Status-Based Prejudice." *American Indian Culture and Research Journal* 15, 3 (1991): 1–17.

Sale, Kirkpatrick. *Conquest of Paradise*. New York: Alfred A. Knopf, 1990.

Sapir, Edward. *Culture, Language, and Personality*. Berkeley: University of California Press, 1956.

Sauer, Carl O. *Agricultural Origins and Dispersals*. Cambridge: M.I.T. Press, 1952. Rpt. 1969.

———. "Maize." *Handbook of South American Indians*. Washington, D.C.: Smithsonian Institution, Bureau of American Ethnology, Bulletin 143: 489–495.

Seaver, James E. *A Narrative of the Life of Mrs. Mary Jemison*. Syracuse: Syracuse University Press, 1975. Rpt. 1990.

Skinner, Linda. "Teaching Through Traditions: Incorporating Native Languages and Cultures into Curricula." *Indian Nations at Risk: Listening to the People*. Washington, D.C.: U.S. Department of Education, 1992.

Sleeter, Christine E., and Carl A. Grant. "An Analysis of Multicultural Education in the United States." *Harvard Educational Review* 57, 4 (November 1987): 421–444.

Smith, Erminnie A. *Myths of the Iroquois*. U.S. Bureau of Ethnology, 2nd Annual Report, 1880–81. Washington, D.C. 1883.

Social Studies (tentative Syllabi) K–12. The University of the State of New York, the State Education Department, Bureau of Curriculum Development. Albany, New York, 1987.

Sosa, John. *The Maya Sky, The Maya World: A Symbolic Analysis of Yucatec Maya Cosmology*. Diss. SUNY/Albany, 1985.

Stuart, George E. "Who Were the 'Mound Builders'?" *National Geographic*, 142, 6, (December 1972), 783–801.

Stensland, Anna Lee. *Literature by and About the American Indian*. Urbana, Illinois: National Council of Teachers of English, 1973.

Stent, Madelon, William R. Hazard, and Harry N. Rivlin. *Cultural Pluralism in Education: A Mandate for Change*. New York: Appleton-Century-Crofts, 1973.

Stocking, George W. "Franz Boas and the Culture Concept in Historical Perspective." *American Anthropologist* 68 (1966): 867–882.

Stone, William. *Life of Joseph Brant—Thayendanegea Including the Indian Wars of the American Revolution.* New York: George Dearborn & Co., 1838.

Sturtevant, E. Lewis. "Indian Corn and the Indian." *The American Naturalist.* Philadelphia: McCalla and Stavely, 1885, 225–234.

———. *Maize: An Attempt at Classification.* New York Agricultural Experiment Station, Geneva, December 1883.

———. *Varieties of Corn.* U.S. Department of Agriculture: Washington, D.C.: Government Printing Office, Bulletin 57, 1899.

———. *Indian Corn.* Thirty-Eighth Annual Report of the New York State Agricultural Society, 1878, Albany, New York: Charles Van Benthuysen & Sons, 1880.

Sun Tracks. *Between Sacred Mountains, Navajo Stories and Lessons from the Land.* Tucson: Sun Tracks, and the University of Arizona Press, 1982.

Szasz, Margaret. *Education and the American Indian, The Road to Self-Determination, 1928–1973.* Albuquerque: University of New Mexico Press, 1974.

———. *Indian Education in the American Colonies, 1607–1783.* Albuquerque: University of New Mexico Press, 1988.

Taylor, Charles. "Interpretation and the Sciences of Man." In Eric Bredo and Walter Feinberg, *Knowledge and Values in Social and Educational Research.* Philadelphia: Temple University Press, 153–186.

Tehanetorens. *Tales of the Iroquois.* Rooseveltown, New York: Akwesasne Notes, 1976.

Teni, Roderico. "Cosmological Importance of Corn among Mayan Peoples." In *Indian Corn of the Americas,* ed. Jose Barreiro, 14–19.

Tooker, Elisabeth. "Ely S. Parker" In *American Indian Intellectuals,* ed. Margot Liberty. St. Paul, Minnesota: West Publishing, 1978, 15–30.

Usner, Daniel, H. "'We Do Not View It So': Iroquois Livelihood in the Face of Jeffersonian Agrarianism." *The John Ben Snow Lecture Series on Iroquois Life and Culture,* ed. Barbara J. Blaszak. Syracuse: Le Moyne College, 1987.

———. *Indians, Settlers, and Slaves in a Frontier Exchange Economy, The Lower Mississippi Valley Before 1783.* Chapel Hill: University of North Carolina Press, 1992.

Vallance, Elizabeth. "Hiding the Hidden Curriculum: An Interpretation of the Language of Justification in Nineteenth-Century Educational Reform." In *The Hidden Curriculum and Moral Education,* Henry Giroux. and David Purpel, eds. Berkeley, California: McCutchan Publishing, 1983, 9–27.

Vecsey, Christopher. *Imagine Ourselves Richly, Mythic Narratives of North American Indians.* New York: Crossroad Publishing, 1988.

Vogel, Virgil J. "The American Indian in American History Textbooks." *Integrated Education* 3 (May–June 1968): 16–32.

Wall, Steve, and Harvey Arden. *WisdomKeepers.* Hillsboro, Oregon: Beyond Words Publishing, Inc., 1990.

Wallace, Henry A., and William L. Brown. *Corn and Its Early Fathers.* Michigan State University Press, 1956.

Wallace, Paul A. W. *The White Roots of Peace.* Philadelphia: University of Pennsylvania Press, 1946. Rpt. Saranac Lake, New York: Chauncy Press, 1986.

Warman, Arturo. "Maize as Organizing Principle: How Corn Shaped Space, Time, and Relationships in the New World." In *Indian Corn of the Americas,* ed. Jose Barreiro, 20–27.

Waugh, F. W. *Iroquois Foods and Food Preparation.* Ottawa, Canada: National Museum of Man, No. 12, Anthropological Series, 1916, Rpt. 1973.

Weatherford, Jack. *Indian Givers—How the Indians of The Americas Transformed the World.* New York: Crown Publishers, 1988.

Wessel, Thomas R. "Agriculture, Indians, and American History." *Agricultural History* 50 (January 1976): 9–20.

Wilson, Gilbert L. *Buffalo Bird Woman's Garden.* St. Paul: Minnesota Historical Society Press, 1917. Rpt. 1987.

Wolcott, Fred R. *Onondaga, Portrait of a Native People.* Syracuse: Syracuse University Press, 1986.

Index